Expert Systems in
Process Control

Expert Systems in Process Control

Fran Jović

Leader,
Design of Process Control Systems;
Professor of Control Engineering,
University of Zagreb and Osijek

and

Section Manager
Computer Systems and Intelligent Process Control
Engineering
ATM Zagreb Company,
Croatia

English Language Edition Consultant:
R.M. Henry,
School of Control Engineering,
University of Bradford

CHAPMAN & HALL
London · Glasgow · New York · Tokyo · Melbourne · Madras

Published by Chapman & Hall, 2–6 Boundary Row, London SE1 8HN

Chapman & Hall, 2–6 Boundary Row, London SE1 8HN, UK

Blackie Academic and Professional, Wester Cleddens Road, Bishopbriggs, Glasgow G64 2NZ, UK

Van Nostrand Reinhold Inc., 115 5th Avenue, New York NY10003, USA

Chapman & Hall Japan, Thomson Publishing Japan, Hirakawacho Nemoto Building, 7F, 1-7-11 Hirakawa-cho, Chiyoda-ku, Tokyo 102, Japan

Chapman & Hall Australia, Thomas Nelson Australia, 102 Dodds Street, South Melbourne, Victoria 3205, Australia

Chapman & Hall India, R. Seshadri, 32 Second Main Road, CIT East, Madras 600 035, India

First edition 1992

© 1992 Fran Jović

Typeset in 10½/12½pt Times by Excel Typesetters Company, Hong Kong
Printed in Great Britain by Hartnolls Ltd., Bodmin, Cornwall.

ISBN 0 412 39730 7   0 442 31457 4 (USA)

A catalogue record for this book is available from the British Library

Library of Congress Cataloging-in-Publication data
Jović, Fran.
    Expert systems in process control/Fran Jović.—1st ed.
        p.   cm.
    Includes bibliographical references and index.
    1. Process control—Data processing.   2. Expert systems (Computer science)—Industrial applications.   I. Title.
    TS156.8.J678   1992
    670.42'75633—dc20   91–40756
                                                                CIP

Contents

Acknowledgements

Our perception or the way we see things makes
all the difference in the world when it comes
to life experiences. It is true that facts do
not create reality. Reality is the way in which
we perceive the facts.

Justin Belitz, O.F.M.
Success: Full Living

The material presented in this book has been evolved from the
PARSYS project on intelligent process instrumentation and parallel
computer systems. The project was initiated by Branko Souček,
Đuro Koruga and Anton Železnikar. Discussions with Petar Brajak
and Sašo Prešern on project material were particularly interesting.
Some theoretical and experimental issues have been the result
of common work with Pavle Marković, Frano Srabotnak, Stjepan
Pliestić and Roman Zupanec. The drawings were designed by
Jadranka Sutlarić. Mira Halar was of much help in English language
consulting. The author would like to thank all of them.

Preface

Advances in artificial intelligence, smart process transmitters and positioners and the overwhelming application of computers in process control – a lucrative and sophisticated technique of the century – have led to the increasing application of expert systems in this field. Three points are still missing: better usability of manpower, handling of the human factor both at customer and system deliverer, and efficient use of process computers. These three aspects are expected to be solved by the end of this century, almost forty years from the first application of an IBM 180 in a Canadian paper mill in 1962.

By pointing out semantic aspects and modelling techniques in expert system application, this book is intended to promote the more efficient use of computers in process control.

The book is divided into four parts

Part One   Concepts, tools and benefits
Part Two   Process surveillance
Part Three Dialogue
Part Four   Action

Chapter 1 gives an overview of intelligent process hardware, presenting the concepts and details of emerging standardization in intelligent transmitter and final control devices. By applying stratified serial buses, intelligent hardware has reached the plant at the very point of field installation.

Chapter 2 tackles the distribution of tools in process control for expert systems that enable coordination and cooperation in solving complex process tasks. Standard features of expert systems in process control are briefly presented.

Chapter 3 gives the fundamentals of benefit calculation when expert systems are specifically applied in process control. The examples originate from power engineering.

Chapter 4 deals with more fundamental aspects of process data generation, presenting three aspects of information: time, space and usage. The calculation stresses the importance of a unified approach to the process data under consideration.

Chapter 5 provides tools for the stratification of process data by making a distinction between process variable states and facts. The importance and reach of these notions and their practical use is illustrated.

Chapter 6 elaborates in more detail the interaction between operator and process, pointing out two aspects of the problem: the designer's presentation and the operator's view of the process. The modern tendency toward open form and dynamic presentation is underlined as crucial in many expert system interactions.

Chapter 7 connects the idea of facts to rules and decision procedures in expert systems. The method of sideways chaining for binary variables is presented. The ordering problem and many valued attributes are also considered.

Chapter 8 gives the operator's standpoint in a dialogue. The limitations and standards are stated. The recognition of process states is preferred.

Chapter 9 makes the distinction between recognition and learning in process control. The procedures of automatic learning and the learning process in operators are illustrated, pointing to the fuzzy facts concept.

Chapter 10 introduces action command and timing in process control, emphasizing basic process states. Common means and procedures for exchanging states to more favoured ones are given.

Chapter 11 elaborates on process protection and automatic actions as opposed to expert system procedures or as limiting factors in the application of expert systems.

Chapter 12 explains the notions of semantics and semantic evaluation of process variables. The connection between process semantics and entropy is also given. The relation between process semantics and action is given as the crucial point in expert system application. The case is given for binary variables.

Chapter 13 treats the original method of modelling stochastic and deterministic linear and non-linear processes. The model is a basis for semantic evaluation of expert system applications. The connection between model variables simulation and fuzzy sets is given. The case of power plant modelling is presented. Significant reductions in the number of rules and time requirements is argued for.

Chapter 14 is devoted to automatic knowledge acquisition in process control. For the case of many-input many-output continuous process variables, an original method of fuzzification is developed. By means of the proposed method a significant improvement in the number of rules and their automatic correction is obtained, yet still enabling a fair time margin for the algorithm. The limits are also discussed.

Thus the book treats the entire cycle of expert systems application in process control, from process data acquisition, presentation, modelling, fuzzification, semantic evaluation and the operator interface to the operationalization of the concept by automatic learning and an on-line algorithm for process control.

The book has been written both as a textbook for students and as a reference for practising engineers and scientists. A minimum undergraduate-level background is assumed and theoretical considerations are explained through practical examples. A list of expert system firms is given. A glossary of terms, an index and reference lists enable further study and practice for all readers.

Fran Jović
Zagreb, Croatia
1991

Glossary

Bayes law	states that the appearance of any event has a certain however small amount of probability
Chaining	infering from the supposed goal rule (backward chaining) or from conditional rule (forward chaining) or by using the priority of the rules according to given process facts (sideways chaining)
Classification	the procedure of attributing different behaviour classes to particular process variables or entities
Confidence function	the individual semantic judgement of the questionable process variable state
Context monitoring	monitoring of specific operator's actions not obviously connected with expected action
Correlation	a systematic relation
Decision	the process and act of choice among many possible ways of behaviour (reaction) or thinking (data processing)
Diagnostic scheme	the usual scheme or plan of actions used for identification of malfunctions on the basis of their manifestation
Entropy	the mean amount of data (in bits) generated by each independent signal character; it is given as the mean logarithmic measure of each signal appearance probability
Evaluation criterion	a standard or measure used for the estimation of system performance
Expert knowledge	the experience of an expert that can be used in building an expert system or solving an actual problem connected with the actual experience of an expert
Expert system	also known as knowledge based system or information based system. These are com-

	puter programs that can mimic many human decision-making processes to give advice, to diagnose and recommend solutions to problems, execute certain critical actions and give explanations
Fact	data triplet composed of an object (physical or conceptual entity), an attribute attached to the object and actual value of the attribute. Thus the flaw value of $3 \, m^3 \, s^{-1}$ at the outlet of the second reservoir of the refinery is a fact that can be proven at any moment with the data recorded in the control system
Fuzzy logic	control logic based on fuzzy sets
Fuzzy sets	sets of attributes described with membership functions whose relations are given with propositions
Heuristics	pertaining to practically acquired knowledge
Inference machine	the part of expert system software that calculates the most probable conclusion from activated process facts using rules, hypotheses, their probabilities and evaluation criteria
Information	the experience of world itself and the environment by a living being
Intelligent hardware	collective term for process instruments, valves and controllers that can change their performance and function using communication facilities process data and supervisory commands
Interaction	the state of mutual or reciprocal action or influence
Interface (technical term)	the place and equipment at which and by which two system elements meet and act on and communicate
Interframe design	the theory and practice of control spaces lay-out with several man-machine interface means such as screens, boards and announciators
Knowledge (machine)	a consistent set of rules probabilities and hypotheses gained by updated process measurements and control actions in the given field of process application
Knowledge base	usually an ordered database containing a

	consistent knowledge of the process under control or surveillance
Large scale system	usually a system with complex interactions with surrounding systems and environment containing a large number of variables and stratified tasks
Learning	the ability of human operator to change his/ her opinion and action because of the acquisition of new process data and information relating to a particular field of learning such as process control
Machine learning	the feature of computer programs for process control to change their actions depending on acquired data on process behaviour and operator instructions by using its own performance function
Modelling	The theory and practice of building various substitutes of actual life or nature without its necessary activation
Operation planning	the usual procedure of long term planning of plant or service development
Ordering of quest	different ways of ordering questions for process facts; because of large number of possible ways a systematic approach should be chosen based on minimum entropy change or minimum uncertainty in the questionning procedure
Performance function	the function that evaluates the performance of system behaviour as the reaction to input, output or control variables; the most common system behaviour is to compensate for changes in process static and dynamic parameters
Problem solving	the technique that integrates problem understanding, devising a plan for solving the problem, carrying out the plan and evaluating it for correctness and for potential as a tool for solving some other problem
Process	a set of operations that perform gradual physical or chemical transformation or a series of transformations; basically material, power and informational processes are distinguished

Process decision	a decision made by operator or intelligent program related to actual or historic process situation or state
Process form	actual state of process and plant presentation; the discrepancy between the presented and presentable is mostly due to designer's and cost efficiency reasons
Process protection	various means used for prevention of the inevitable effect of process state instability
Process state	a distinct state of crucial process variables that is the result of internal and external process or plant forces or actions
Process variable	a real or from real data calculated row of data or time series usually expressed in engineering units representing certain process value such as flow, level or temperature
Production rule	expert system rule connecting input output and control variables in a systematic way
Reaction	the usual result of process decision; such as in operators when process change takes place
Recipe logic	deterministic production rules based on heuristics and usually not of much value for problem solving
Recognition	the ability of process operator to resolve among particular process states those states crucial for future process behaviour
Rule	the basic unit of an expert system. It consists of one or more conditions and one or more conclusions under given hypotheses
Rule value	a relative weight of a rule based on probabilistic calculation of all hypotheses when certain facts are 'activated'
Semantic	several attributes that can be connected with the state of process variables that possess certain meaning to particular process operator or superintendant; these attributes are existence, relevance and importance
Semantic parameter	analytical measure used for the evaluation of meaning, communicated by process data; such measure is inevitably individual
Simulation	the procedure of model activation verification and modification
System	a combination of units as to form a whole

	and to operate in a unison executing a given function or a set of function; essential for any system is the definition of its environment and performance measure
Symbolic model	the communication of most complex objects and their structures by audio-visual means by a properly designed interframe

Part One

Concepts, Tools and Benefits

Three basically different approaches to expert system applications in process control are described. These concepts are evaluated from the standpoint of process information reversibility. Standard tools and their efficiency are described and analysed. The complex preconditions for this degree of automation are scrutinized. A brief survey of the expected tangible and intangible benefits is given.

1
The intelligent hardware concept

1.1 INTRODUCTION

Compared to classical measurement transmitters, an intelligent instrument is a device based on a microprocessor which allows (Figure 1.1):

1. greater accuracy of measurement through compensation for temperature changes and systematic errors;
2. a higher measurement range;
3. a communication interface to the operator or maintenance staff through a hand-held communicator or computer interface.

By using an intelligent, or 'smart', transmitter and positioner the whole process loop becomes more flexible, accurate and operator interactive than in standard systems. The intelligent transmitter produces both digital and analogue data (e.g. 4–20 mA) making the interface to the operator self-calibrating and extending measurement ranges without the need for it to be dismantled and mechanically manipulated and changed. The features of intelligent positioners are similar, allowing the flow characteristic curve to be changed, such as from linear to equal percentage (Figure. 1.2) and various control features to be changed (Figure 1.3). Intelligent behaviour is also an expected feature of control computer software, and is the most important aspect of the interface both to the process devices and to other computers in a network. Thus the intelligent communication feature among all process devices provides a basis for intelligent process control. Projects neglecting this fact are prone to failure. The interfacing and standardization of intelligent devices are still in much dispute and controversy. Therefore the interfacing philosophy will be presented in more detail.

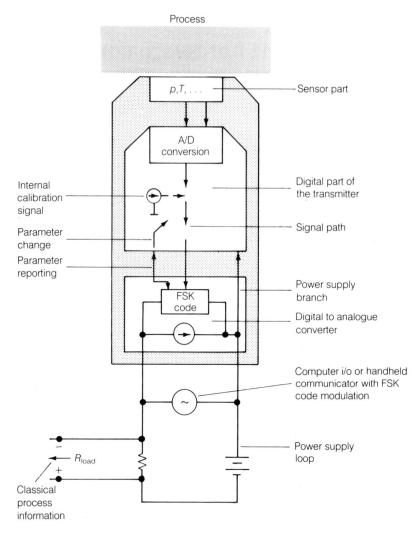

Figure 1.1 (*a*) Block scheme of an intelligent pressure transmitter and its communications; (*b*) block diagram of the Rosemount 3051 intelligent pressure transmitter.

1.2   INTERFACING, STRATIFICATION AND THE BASIC PROCESS UNIT

A process control system is usually distributed in the plant (see Figure 1.4) as a two- or three-level hierarchical assembly and connected via serial redundant buses to some kind of operator's station. Further architecture involves some supervisory and plant managerial

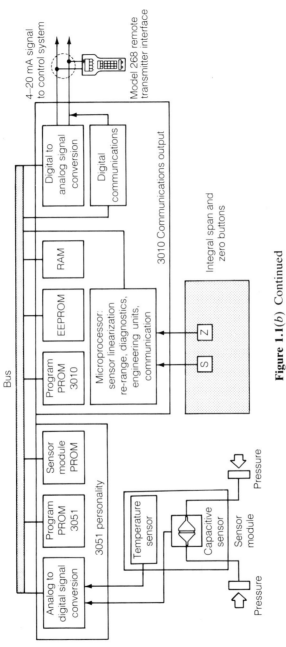

Figure 1.1(*b*) Continued

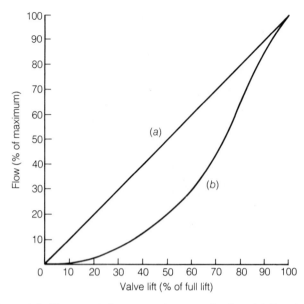

Figure 1.2 Characteristic curves of a control valve: (*a*) flow is directly proportional to lift; (*b*) flow changes by a constant percentage of its instantaneous value for each unit change in lift.

computers where historic records and expert system tools can be implemented. The stratification procedure is crucial to the cost-effectiveness of the plant since it sometimes involves a tricky trade-off between the high software cost of large process unit computers and the high communication cost for a fast data exchange among a greater number of basic process units. There is a typical solution trade-off. The variants, however, depend on the controlled process. Some processes demand large and sophisticated process computer applications, especially for fast and potentially dangerous processes. The interface between operator and computer demands special treatment, since it involves the problem of responsibility and process integrity. In order to respond to process state events, the operator has to have a very large database. Accumulating huge amounts of data, for example of the order of several Gbytes per day, the problem of their refinement, sense and relevance appears. Many procedures have been proposed for the reduction of data in process control in order to interface the operator and the process properly. Some of them, such as entropy, and semantic and expert system concepts, will be elaborated here. However, the issue of so-called

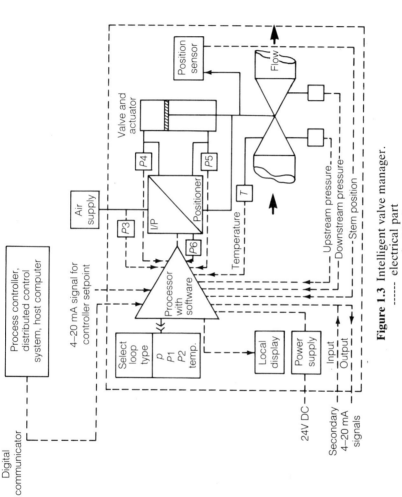

Figure 1.3 Intelligent valve manager.
------ electrical part
——— non-electrical part

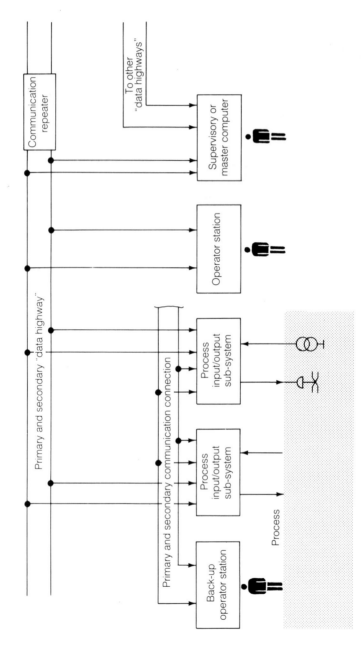

Figure 1.4 Distributive intelligent input−output sub-system − a system integration concept.

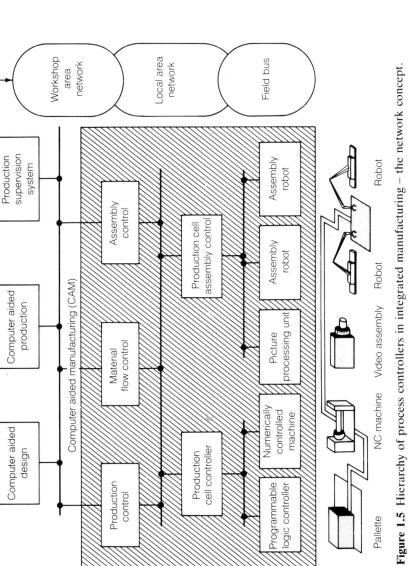

Figure 1.5 Hierarchy of process controllers in integrated manufacturing – the network concept.

Table 1.1 Some MAP compatible bus interconnections for process control

MAP level	Function	MAP structured bus	FIP bus**	PROFIBUS***	State
7	Application	RS 511-Mini MAP	FIP application	PROFIBUS application	Actual
6	Presentation	–	–	–	Empty
5	Session	–	–	–	Empty
4	Transport	–	–	–	Empty
3	Network	802.1* (address & architecture)	–	–	Partly used
2	Data link	LLC 802.2*	FIP application	PROFIBUS with limited token passing IEC TC57 (FT1.2)	Actual
1	Physical	802.4 token bus fast field bus slow field bus	FIP transfer protocol	RS 485 EEx RS 485 isolated total field bus management	Actual

* CCITT standards
** Common French-European industry project
*** Common German-European industry project

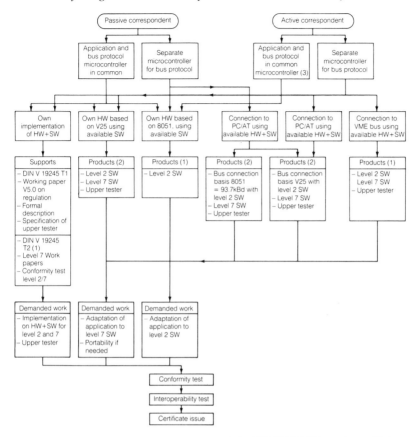

Figure 1.6 Different realization schemes of the PROFIBUS interface: (1) in preparation, (2) products at disposal, (3) depending on processor power and application complexity.

machine intelligence evaluation is only briefly considered in this book.

1.3 INTERFACING STANDARDS –
THE FAMOUS PROFIBUS AND FIELD BUS

Interfacing production machines and various process control computers demands standardization of communications (Figure 1.5). Interfacing operators and machines in an efficient manner is another problem and requires intelligent correspondents – part of this will be elaborated in this book (Chapters 9 and 14), and part of it is the problem of artificial intelligence, hence out of the scope of this book

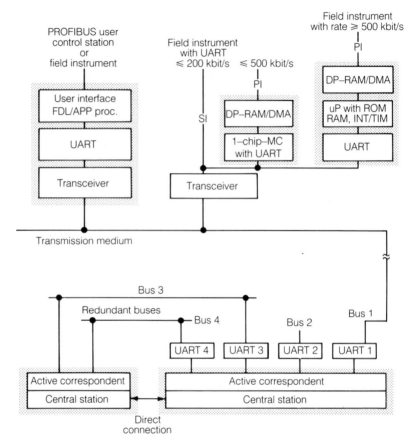

Figure 1.7 Different PROFIBUS connections. SI: serial interface; PI: parallel interface; DP-RAM: dual port RAM; DMA: direct memory access.

(Science News, May 15). Today Man Machine Interface (MMI) is a more or less rigid definition of maps, lists, menus and protocols as a result of its historic development and does not suit a sophisticated user. Especially unfavourable is its price/performance ratio. Some improvements in this field are presented in Parts Two and Three.

Interfacing machines in process control means connecting transmitters, controllers, actuators, auxilary control equipment, supervising computers and factory management systems into a useful system that dynamically changes its function and goals. There are many approaches to this problem, (some of them given in Table 1.1) and most of them based on MAP standards (Kazahaya, 1987) of the OSI open system concept. As the most advanced manufacturer-

Table 1.2 Electrical and mechanical characteristics of the PROFIBUS physical link (according to DIN V 19245 T1)

Bus structure	: Terminated and open-end lines with branches
Transmission medium, length, no. of participants	: Dependent on transmission technique, i.e. twisted pair line with grounding protection, ≤ 1.2 km without amplifier, 32 participants
Redundancy	: Optional second medium
Station types	: Active participants with access control Passive participants without access control
Address distribution	: 0–127; 6 bit address extension/segmentation
Transmission method	: Half duplex, asynchronous, burst free synchronization
Bus access	: Hybrid, decentral/central token passing between active participants and master–slave between active and passive participants
Transmission services	: Acyclic: Send Data with no Acknowledge Request Data with Reply Send and Request Data Polling: Request Data with Reply Send and Request Data
Message length	: 1.3–255 byte per telegram, 0–246 byte net to data (Level 2) per telegram without address extension
Transmission	: Telegrams with Hamming distance (Hd) = 2 or 4, Burst recognition sequences against loss and duplication of messages

supported solution, the PROFIBUS interface is presented in more detail (DIN V 19245 T1) in Figure 1.6. The structure of PROFIBUS connection variants is given in Figure 1.7. Electrical and system details are given in Table 1.2.

1.4 INTELLIGENT TRANSMITTERS AND POSITIONERS: THE HONEYWELL CASE

Ninety-five per cent of control loops are closed loops. Thus, controlling a temperature requires a set-up like the one in Figure 1.8. Putting 'intelligence' into this loop consists of:

1. making the temperature sensor and control device with digitally exchangeable features and self-diagnostics;
2. connecting them into a communication set-up;

Table 1.3 Comparative data on three intelligent transmitters

	Rosemount	Honeywell	Sattcontrol
Transmitter model	3051 C	ST 3000	The GOLD series
Measurement type	Differential pressure	Differential pressure	Differential pressure
Measurement ranges (kPa)	0–0.2 I 0–6.22 0–2.07 II 0–62.2 0–8.28 III 0–248	0–1 inch H_2O 0–400 inch H_2O 0–5 PSI 0–100 PSI 0–100 PSI 0–3000 PSI	0–2 KPa min 0–190 KPa max
Input resolution	N.A.	N.A.	20 bit or 0.0001%
Turndown ratio	30:1	From 20:1 to 400:1	75:1
Guaranteed accuracy (%) of calibrated span	±0.10% for spans from 1:1 to 15:1	0.10%	0.10%
Communication link	HART protocol: BELL 202 Frequency Shift Keying (FSK) technique	Through existing 4–20 mA loop or TDC digital communication	HPIL (Hewlett Packard Instrumentation Link)
Self-calibration & Temperature compensation & self-diagnostics	Yes	Yes	Yes

3. enabling operator intervention through some type of smart communication device.

This gives the following benefits:

1. accuracy and stability increase;
2. process control functions (for instance linear, percentage or non-linear flow control) can be changed without hardware intervention;
3. maintenance cost drop of up to 20 per cent compared to the conventional process devices.

Three smart transmitter products are compared in Table 1.3.

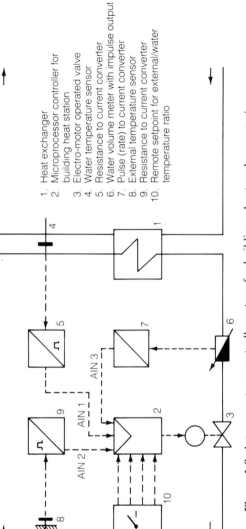

1. Heat exchanger
2. Microprocessor controller for building heat station
3. Electro-motor operated valve
4. Water temperature sensor
5. Resistance to current converter
6. Water volume meter with impulse output
7. Pulse (rate) to current converter
8. External temperature sensor
9. Resistance to current converter
10. Remote setpoint for external/water temperature ratio

Figure 1.8 A temperature controller set-up for building a heat exchange station.

2
Distribution and cooperation – expert tools

With its intelligent data processing, distributed process control penetrates into practically each process device. However, in taking only measurable data and connecting them in a network structure an important part of reality is neglected: the process itself. Consisting mainly of plant elements and media, the process needs informative integration into the system. Only the integration of all plant parts and devices makes an entity that allows cooperation among the plant parts and their surroundings. Classically, it was the duty of trained operators and maintenance staff to make up for this lack of information completeness. Up to now only expert systems have shown the necessity to undertake this vital task. In this way a unique feature of integrated process control is established.

Application approaches to expert systems in process control differ according to the design milieu. Large companies, like ABB (ASEA–Brown–Boveri), prefer a more complex approach with its GRADIENT project. Specialized companies, such as Gensym, offer commercial expert system tools (e.g. Gensym's G2) and researchers from universities develop specific software tools for each application, such as SOLEIL at Hahn–Meitner Institute, Germany. The ABB project named GRADIENT (1990) has two main goals

1. research of the knowledge base as an operator support tool for the industrial control of large and complex objects such as a power plant;
2. interfacing the operator with the process control system by means of a graphic expert system.

These goals can be achieved by using a set-up of several expert systems or by cooperative expert systems. Such systems serve as a consultant to the process operator, as a control system and partly as a dialogue system. Thus they help operators during a failure, during plant work modification, during abnormal work conditions and dur-

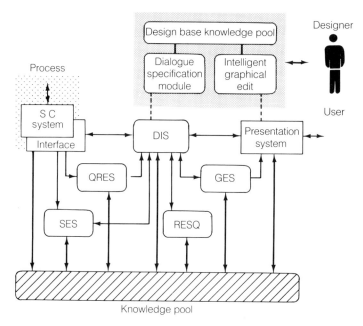

Figure 2.1 The expert system GRADIENT. SES: support expert system;
QRES: quick response system; RESQ: operator response evaluation system;
DIS: dialogue system; GES: graphical expert system.

ing normal plant operation. Components of the system GRADIENT
are shown in Figure 2.1. The assessment of the system state, the
estimation and assessment of the actual goal state and the estimation
of the operator state are essential for the behaviour of the expert
system.

The real-time expert system G2 (Figure 2.2) is a commerical
product of Sira Ltd, England, intended for medium to large processes with hundreds or thousands of variables. Interactive graphics
and structured natural language are used for direct creation and
control of the knowledge base. The knowledge is presented schematically, by dynamic models and heuristic data. Real process
objects, their connections and attributes are also presented as
schema. Dynamic models can be attached to the schematized objects
in the form of a schematic heuristic or analytic presentation. These
models can be used in the knowledge base for simulations or for
the comparison of the expected and real behaviour of the object.
Heuristic data are expressed in knowledge frames containing rules,

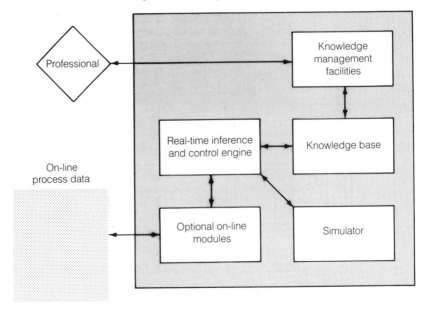

Figure 2.2 G2: direct building and management of a real-time expert system including simulation testing of the knowledge base prior to on-line use.

problem type and referred objects. The expert system behaviour is determined by an inference machine (program) which calls and evaluates relevant rules.

2.1 THE PLACE OF EXPERT TOOLS IN PROCESS CONTROL

Expert tools are used to support the usual process control set-up with the following tasks:

1. to ensure tangible benefits, such as better productivity in power plants;
2. to provide operators and plant management with better insight into the 'hidden' processes in the plant;
3. to allow better operation planning;
4. to preserve and enhance knowledge of the type 'know-how'/ 'know-why' of the tecnical staff;
5. to enhance the working motivation of the technical staff.

Therefore they are installed at the operators' desks, at the operational management positions, at expert personal computers and in the service area. Successful expert tools are rare because of some common reasons such as:

1. lack of a real process expert;
2. neglecting the existing expert knowledge and experience;
3. inadequate hardware and software installed at the site;
4. the process allows no additional overhead or is too complex or interdependent or some input ingredients are unknown/undetectable;
5. inadequate interest from the owners.

There are many inadequacies of expert system tools in process control:

1. a better performance by classical process control than by so-called 'recipe logic';
2. 'production rules' that neglect process facts;
3. diagnostic schemes that do not solve a failure;
4. process models and simulation lacking connection to expert tools;
5. a 'fuzzy set' concept lacking even a minimum of dialectics.

Besides a general lack of performance of expert systems, deficiencies also exists in the following fields:

1. planning – where computers tend to calculate all possible moves, which is often unmanageable;
2. reasoning – where the lack of criteria does not allow the abandoning of 'irrational' rules or the selection of the most 'rational' conclusion;
3. learning – where the precision of statistical data on which the inference is performed is not known in advance by a machine

2.2 THE TOOLS: G2

As pointed out above, approaches to expert system applications differ according to the design milieu. Roughly speaking, major process control equipment suppliers seek solutions for large-scale control systems such as power plants, while specialized expert system companies support an all-round solution. Research teams at non-profit-making institutions undertake complex process control research in expert system tools and methodology.

The Gensym G2 real-time expert system is designed for large applications where hundreds or thousands of variables are monitored concurrently. G2 is tailored for such complex real-time applications as the process control, computer integrated manufacturing, network monitoring and automatic testing. G2 uses schematics, dynamic models and heuristics to represent the professional's knowledge of the application.

Schematics are collections of objects, frames and connections that represent real-world objects and their relations to other objects. Schematics can be viewed at various magnifications for either an overall or detailed view. Windows containing information about objects can be brought up for editing of attributes or models of these objects. The expert can define new objects, define attributes and graphically edit the related schemes.

Dynamic models can be attached to objects. They may be the heuristic or analytic representation of behaviour over time. These models can be used for simulation to test knowledge bases and for on-line use to compare expected with observed behaviour.

Heuristics are defined in knowledge frames that include not only a rule, but also associated information, such as the problem type and objects to which the rule applies. Rules are in generic form: that is, they apply to all objects of a given type. Further, rules may include references to connected objects through multiple levels. The schematic provides the basis for interpreting a rule as it applies to specific objects and to objects connected to it.

Historical data can be referred to by rules and analytical data. Time-based properties of variables, such as rate of change and time series statistics can be expressed in structured natural language.

Real-time procedures give G2 the ability to carry out many different plans or strategies concurrently. These procedures may be executed over extended periods of time and give the system an attention span beyond the present moment.

2.2.1 KNOWLEDGE MANAGEMENT IN G2

Structured natural language commands enable the user to create knowledge. The syntax and grammar are designed for real-time applications and the grammar can be customized to specific application domains. The system interactively guides the expert in entering and editing knowledge. As the professional user enters and edits knowledge, menu windows pop up to guide the user through the structured natural language input. Each knowledge frame can be rapidly created and fully integrated – in a compiled form – into the knowledge base. Knowledge may be edited interactively, even while running on-line.

Dynamic simulation permits the user to test the knowledge prior to on-line use. Failure modes, disturbances and dynamic behaviour can be represented by heuristic or analytic models, including non-linear differential equations and logic expressions. Simulation testing is animated through dynamic meters, graphics and charts which can be interactively defined and manipulated.

Explanations of reasoning can be requested or automatically produced. Explanations show the rules involved in deciding that a particular action – such as generating a message – should be taken, as well as the values of variables referred to in these rules. The values of variables are typically shown as trend graphs that plot several variables against time.

Structured natural language queries allow the user to retrieve knowledge on the basis of attributes, such as the problem type, authors, object type, or object name. Related knowledge can be examined in a single workspace to review completeness and consistency and to allow interactive editing.

2.2.2 INFERENCE IN G2

Peripheral awareness of a significant behaviour can be maintained by repeated testing for specific conditions or patterns of conditions. Such repeated testing of scanning can either be done by G2 directly or can be delegated to lower-level data servers that can alert G2 as appropriate.

The focus mechanism makes use of metaknowledge, such as the problem type and object type, to determine which knowledge frames to invoke. The focus is augmented by conventional forward and backward chaining. Multiple problem areas can be focused concurrently.

Truth maintenance considers time duration as well as changes in data. Data are tagged with a time-stamp and validity interval. All inferences and analytic calculation based on data reflect this to ensure the current validity of high-level conclusions.

A real-time scheduler manages the tasks intrinsic to G2.

The working environment is presented in Figure 2.3 and a table-object and associated data in Figure 2.4.

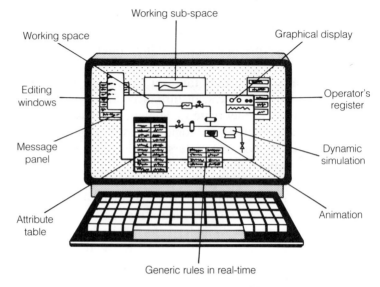

Figure 2.3 Working environment in the G2 expert system.

For industrial process control the crucial point is to present the right information in the right way at the right time. In GRADIENT the knowledge-based systems provide the content of the interaction, the graphical expert systems provide the form, and the whole is controlled by a dialogue system. There are, accordingly, two user categories for the GRADIENT system. One is the designer of a particular application, and the other is the operator who uses it to control the process.

2.2.3 SUPPORT FOR THE DESIGNER

It is a basic assumption of the GRADIENT project that it is not feasible for the designer to specify the actual interaction in every detail or to account for every possible contingency. The designer should rather provide the overall strategies and principles of dialogue control that are to be applied. For this purpose the designer needs a number of powerful tools – for dialogue specification, for the knowledge representation and for the graphical design which enables the designer to develop and implement a specific application from the general functionality provided by these tools, and to structure the form and content of the dialogue.

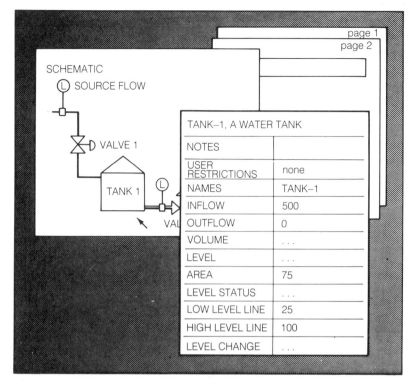

Figure 2.4 A table of object-associated data in a G2 expert system.

2.2.4 SUPPORT FOR THE OPERATOR

In the GRADIENT project the emphasis has been put on facilitating the response to a disturbance through intelligent alarm handling and the provision of recommendations and on reducing the occurrence of 'human errors' through context-sensitive monitoring of actions. The output of the specific knowledge-based systems constitutes the main content of the dialogue. This is presented to the operator according to the specifications provided by the designer, as implemented by the functionality in the dialogue system and the graphical expert system.

2.2.5 THE GRADIENT WORKPLAN

The objectives of the GRADIENT project are achieved through an overall system development based on a representative example of a process control system. In GRADIENT this is realized by the implementation of two process simulators being:

1.   a conventional power plant, and
2.   a data (packet switching) network

The processes are available in the form of computer simulation, with the modifications necessary for sufficient experimental control. Some basic functions of the supervisory and control system and the presentation system necessary for the process models are also developed and implemented in the simulator systems. The simulators serve as vehicles for the test and evaluation of the individual GRADIENT systems.

2.2.6   GRADIENT SYSTEM COMPONENTS

Quick response system (QRES)
The purpose of the QRES is to support the operator during a system failure. In order to accomplish this, the QRES must identify system failures and system failure states, inform the operator of their existence, evaluate their severity and recommend the proper corrective action. This task includes continuous supervision of a selected (representative and significant) part of the dynamic process and a reporting of those failure states that must be brought to the attention of the operator, together with a preliminary recommendation.

Support expert system (SES)
According to the current state of the art of ergonomics, in a failure, the following operator actions are important:

1.   detect the failure (QRES),
2.   stabilize the process (SES procedural support),
3.   evaluate the consequences (SES consequences prediction), and
4.   investigate the causes (SES state-based diagnosis)

Procedural support
Procedural support serves to help the operator in handling the process. It entails reasoning about actual goals to define the sequence of sub-goals – and thereby the actions or an action envelope – that can achieve the intended result, e.g. bring the system to the desired (goal) state.

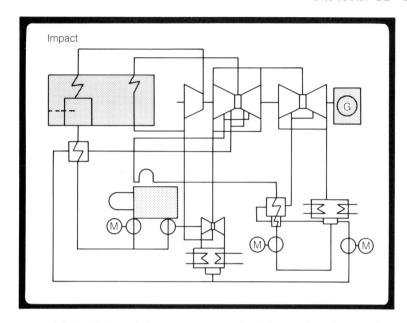

Figure 2.5 Prediction of the consequences of a failure and explanation by SES.

Prediction of consequences of failures

The SES will predict the consequences of failures to support the operator in his task of choosing correct actions to compensate for the effect of one or more failure states in the process. The failures can either be actual failures as signalled from the Supervision and Control System failure states detected by the QRES module or hypothetical failures indicated by the operator.

State-based diagnostics of causes of failures

The SES will also support the operator during the diagnosis of failures. After the QRES has detected a failure state and the operator has stabilized the process (using the QRES recommendations and possibly the SES procedural support) the SES can be asked for a detailed diagnosis of the causes. The SES state-based diagnosis can use information queried from the operator in addition to the information provided by the supervisory and control system. SES can also test hypotheses about a possible failure situation provided by the operator.

Figure 2.5 presents the prediction of consequences of a failure and explanations by the SES.

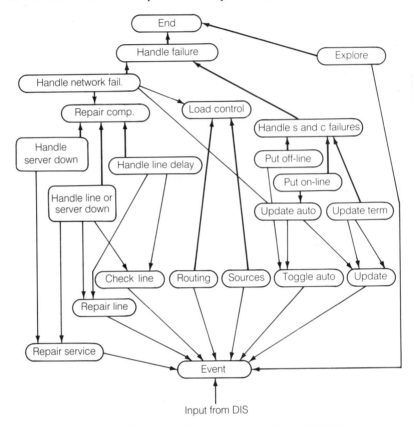

Figure 2.6 Hierarchy of plans and sub-plans in RESQ.

Response evaluation system (RESQ)
The RESQ monitors the operator's actions to detected situations
where the pattern is inconsistent or incorrect. Such situations occur,
for example, when operators forget actions that were taken some
time ago, when the execution of a action has been delayed as several
lines of action are executed simultaneously or in the transition
between work shifts.

The RESQ is based on the combined use of the techniques of plan
recognition and detection of erroneous actions. The RESQ analyses
the actions of the supervisory and control system operator, attempts
to recognize what plans and goals the operator is trying to achieve
and evalutes these goal in terms of the current system status. The
structure of the RESQ is given in Figure 2.6.

Dialogue system (DIS)
The dialogue system has a central role within the GRADIENT system and has two principal functions:

1. it channels communications from the SES, QRES and RESQ to the operator(s) and handles the operator queries to the SES, and
2. it provides the content to such exchanges from the dialogue specification contained in its dialogue assistants.

The DIS does not handle the 'front-end' dialogue, which, in the form of input/output, is the job of the presentation system. The process information is monitored by the dialogue system but is routed directly to the presentation system for display.

Graphical expert system (GES)
The GES is intended to support the dialogue system by handling the main stream of measurement and control information from the supervisory and control system. The GES will dynamically compose pictures and picture sequences using knowledge of the process, the user, graphical representation model techniques and dialogue techniques. The main emphasis is put on the graphics-orientated functionality, i.e. creating, influencing and modifying graphical representations on-line. The GES is preceded by several phases of Intelligent Graphical Editor (IGE): designer support for engineering-orientated tasks and semi-automatic picture generation from the underlying process-related knowledge representations, design and prototyping of graphical operator dialogues and ergonomic and user modelling aspects.

The presentation system (PRES)
The PRES is an interface between the operator and a given dynamic system. It allows the operator to get the information required by acting as a delivery system for graphical specifications generated through the Intelligent Graphical Editor (IGE). These presentation descriptions are subsequently driven by the dialogue system through dialogue specifications held in its dialogue assistants. The PRES also allows the operator to make changes in the process states.

3
Benefits and metrication

The crucial point in the expert system application is the benefits it can bring. One cannot expect a single-valued answer to this question. But by examining the production efficiency between the first, second and third shifts in a power plant we can get data like those in Figure 3.1. Long-range control of these differences can be the reason for the application of the expert system in reducing the effects of the human factor. Such benefits can be quantified. By comparing the cost of fuel with the same productivity of power for a concrete case one can expect benefits in the range of one per cent. For a small power plant of the range of 100 tons of steam per hour this can mean a saving of US$ 200 000 a year.

3.1 TANGIBLE BENEFITS

Some preconditions should be obtained prior to the application of the expert system:

1. minimum willingness of the process expert to cooperate for the given process part;
2. minimum installed equipment for the measurement, control and supervision that gives initial data for a preliminary or feasibility study for the application of the expert system – one such system installed at a 30 MW power plant is given in Figure 3.2 provided by ATM – Zagreb (Anić *et al.*, 1989)
3. process conditions that can support additional expenses that exhibit a satisfactory but not excessive complexity and interdependency of parts – to some extent the contents and composition of materials and power supplies must be measurable and controllable;
4. adequate interest of the plant owner in the improved operation and productivity of the process.

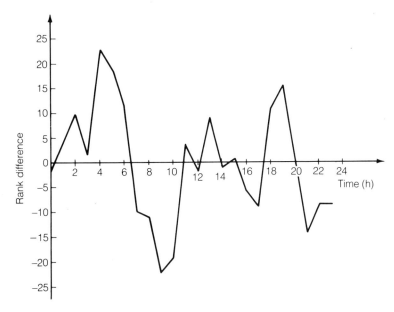

Figure 3.1 Cycle of daily rank difference of the main input–output variables of the thermal power plant.

Without the fulfilment of all these conditions there is no point in starting the expert control task.

The most common tangible benefits are: extension of the life of the equipment, extension of the dynamic margins of the plant, better product quality and better service quality.

Extension of the life of the equipment is obtained by:

1. earlier detection of failures, preventing the spread of failure effects;
2. better tuning of the system to dynamic changes in the process (see Chapter 13);
3. preventive modelling of process elements responsible for damage prior to their full appearance.

The extension of the dynamic margins of the plant is obtained by:

1. using the plant in the second and third shifts at the same quality as in the first;
2. enabling higher plant dynamics by knowing how to prevent otherwise unavoidable failures;

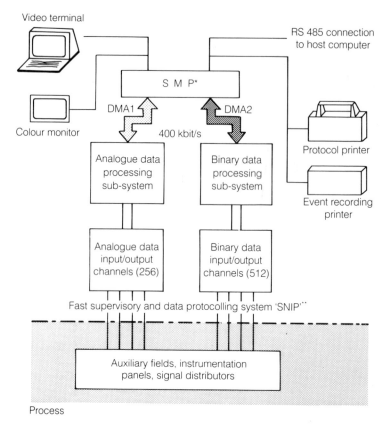

Figure 3.2 Fast data accuisition and protocolling system for supervisory control
of a 30 MW power plant.
* Siemens modular system based on Intel family of microprocessors.
** trademark of ATM-Zagreb.

3.   using less qualitative power or material or service resources
     while making an expert adaptation to such a situation.

A better product quality is obtained by:

1.   better control of the production process;
2.   better control of the critical process parameters;
3.   obeying special rules that cover the issue of the product quality;
4.   better response of the system to changes in relevant material
     and/or power components or services responsible for the product
     quality.

Better service quality is obtained as a result of:

1. better control of service parameters, such as heating water temperature, or burner gas pressure;
2. dynamic adaptation of service to external conditions such as weather, time of day or day of the week;
3. faster response to problems in the service departments;
4. fewer problems in service activities.

Figure 3.1 presents rank differences for a power plant's main input and output variables, fuel versus superheated steam, where the main rank differences occur at the end of the third shift and during the early hours of the first shift. Total losses were estimated as 1.5%. By applying the control algorithm as given in Chapter 13 the control settings were close to reality with 5% accuracy (Marković, 1990). Thus benefits of 1% can be expected in this case.

3.2 INTANGIBLE BENEFITS

Among the main intangible benefits are: retention of expertise, better planning, easier work for operators and technologists, saving and improvement of the owner's technology and higher working motivation for experts.

3.3 BENEFITS, CONFIDENCE AND WEAK POINTS

Expert systems are based on process heuristics obtained either by the process or by automatic data acquisition and data processing. Such heuristics are subject to modifications during plant operation. Modifications are structural when global process changes are made and are related to a different method of plant functioning and a completely different plant rule description. Qualitative changes induce changes of rules that describe process behaviour. An example of a qualitative change is when a process input change of 3% is regarded as a small change compared with the previous process state, where the same change was regarded as a medium change. The other problem of process heuristics is the quantitative evaluation of process changes and states. There are many ways of quantitatively evaluating process heuristics, such as:

1. *Bayes' Law*: the values of process variables are quantified according to heuristically obtained data, such as probabilities or expectations.
2. Macro modelling: heuristically defined global process variables are interrelated in quantitative model relations that serve for prediction, feed-forward or adaptive control.
3. Markov chains: similar to Bayes' Law; heuristics define significant process states depending on the process input and the previous process state. Process transition into the next state is defined with the transition probability and renewed with actual process measurements.
4. Fuzzy sets: process variables are defined heuristically as well as their changes and their membership functions. Several process variables cover the same value of the process variable or its change.
5. Neural network: a three-layer neuron structure imitates the neuron transfer function – it serves for the adaptation to the expected response to the process input after a learning period which demands a large set of examples.

In this book, macro modelling is applied to the semantic calculation of the process, Markov chains to the state modelling of the process, and modified hierarchical fuzzy sets to better process control tuning. Unless the system provides a fair approach based on obeying the Bayes estimation and/or heuristical data application, no one can expect benefits. Confidence in the system results from the heuristics and Bayes' Law approach and from long-term tuning of process model data (see Chapter 14).

3.4   METRICATION

In the quest for expert system metrication, many companies and international projects have made substantial efforts, such as ABB and ESPRIT (ESPRIT Project P 857). This is a fundamental question regarding the productivity of these projects and programs and a practical point of the adequacy of presentation and use of knowledge engineering.

The reversibility rate of data processing can be stated as a reasonable measure for the evelution of process control systems and expert systems as well. In fact, the majority of process data are destructively overwritten at each cycle of data acquisition, without any

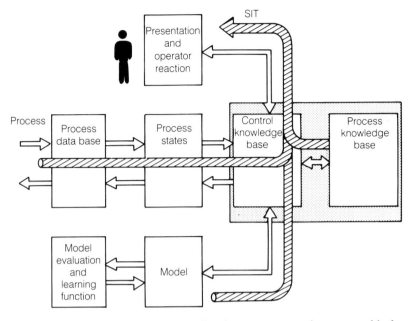

Figure 3.3 A simplified semantic data flow in process control system usable for semantic calculation.

possibility of their reconstruction. This prevents the reversibility of process observations. In order to calculate the reversibility rationally, one has to forget the information ratio in bits, as stated by Shannon, and start to calculate the semantic parameters of process variables (see Chapter 12), evaluated as the SIT (or semantic unit of information flow). Therefore, to each process variable at every time (instant) of data collection, a semantic value can be attached, i.e. for each variable there stands a relation

$$Q_{i,n} = (e_{i,n}, q_{i,n}, r_{i,n}) \qquad (3.1)$$

where $Q_{i,n}$ is the quality of process information of the variable v_i, $q_{i,n}$ is the relevance of variable v_i, $e_{i,n}$ is the existence parameter of variable v_i and $r_{i,n}$ is the importance of variable v_i all at the nth time instant. A more precise calculation of $Q_{i,n}$ is given in Chapter 12. Since there is subjectivity in each semantic evaluation, the quantity of semantic data can be different for each process control participant. Nevertheless, because of the very small amount of data needed to follow each semantic situation in a process and the one-dimensionality of the registration, this can be effectively reproduced. Figure 3.3 shows a pictorial method of semantic content calculation. Comparing this figure with the data acquisition schemes given in Figures 2.1 and 2.2 for different expert systems one can conclude:

1. the GRADIENT system, with the included model of the power block preserves the greatest resemblance to the representation given in Figure 3.3; the greatest attention of the operator is dedicated to the operator reaction to the semantic data flow.
2. the G2 system, without additional model data, can be considered successful only when pointing to relevant semantic data (i.e. facts); for a higher number of complex variables without model and simulation tools it could be questionable.
3. the SOLEIL system (SOLEIL 1989) works on the basis of process state identification, which can usually lead to the destruction of relevant semantic data, since the process state description is not explicitly connected to the semantic content.

REFERENCES FOR PART ONE

ABB (1989) GRADIENT – Graphical Dialogue Environment. Graphical and knowledge-based dialogue in dynamic systems. ABB Publication No. D CRH 1333 89 E.

Amato, I. (1990) Inventing life, *Science News*, **137**(20), 312–14.

Anić-Ivičić, S. *et al.* (1988) *SNIP-Supervisory and plant protocoling system*. Proceedings of the seventh Yugoslav Symposium on Microprocessor Application in Process Control. MIPRO-PU, Rijeka, 5–1/5–5.

Gensym Corporation (1989) *An introduction to G2*.

Gitt, W. (1986) *Energie-optimal durch Information*, Hšnssler, Neuhausen-Stuttgart.

Hahn-Meitner Institute (1989) *SOLEIL, ein Expertensystem zur Steuerung der Plasmandepositionsanlage*.

Jović, F. (1988) *Kompjutersko vođenje procesa*. ZOTKS, Ljubljana.

Jović, F. and Zupanec, R. (1990) Proc. of the 35th Annual Gathering, Jurema, Zagreb, Jurema, 122.

Kazahaya, M. (1987) Field bus: new standard. *Control Engineering*, 2 October, 21–4.

Marković, P. (1990) *Fuzzy set hierarchical model with variable structure*. Ph.D. thesis, Faculty of Electrotechnics, University of Zagreb.

Part Two

Process Surveillance

Process control systems are developed, designed, used, maintained, repaired and abandoned, or redeveloped, by and for humans. Production processes are therefore a series of interactions: process–control system–human–control system–process. The integration of humans into the production process should be planned, adequately investigated, supported and developed.

Certain characteristic human roles are recognized in the system usage: system operator(s), process operator(s), maintenance staff and training staff. The role of humans in process control functions in production (i.e. process operator or dispatcher functions and facilities) is emphasized.

More efficient use of process computers is to be expected by the end of this century (Macrae, 1988). Thus, from the first installed process computer, an IBM 180 in a Canadian paper mill in 1962, there has been no essential change in favour of the user. Reasons are human factors in the user, human factors in the supplier and a low level of usability of the equipment for the process operator. The intention is to present methods, ways and means for a better usability of process computers. But to cite Pauline Kael 'Surely there are no hard and fast rules: it all depends on how it's done'.

The function of a process operator or dispatcher is to manage the system, keeping it in balance, and to produce the product at the right time in the correct quantity and in an appropriate and safe environment for both humans and equipment. The operator or dispatcher has numerous subsidiary managerial roles communicating with other humans involved in the process, controlling various process states and occasionally reporting or consulting other services and teams. The basic task is to make decisions on production that can be supported by:

1. sufficient information on everything operating normally or on places where abnormalities are present;
2. certain records of events that have occurred, knowing where the equipment is installed, which devices have obtained commands, which part of the process is supposed to be in a certain state;
3. information on the process variable states and trends to determine whether the process is working properly;
4. facilities to issue necessary commands and information on the results of command execution.

Additional features, as listed below, can help the operator or dispatcher perform his/her function in the best way:

1. Fast reaction of the control equipment and a real feeling of being present at the process site;
2. Minimum noise in the control room;
3. Appropriate seating;
4. Proper lighting;
5. Minimum outside interferences, especially by telephone;
6. Moderate or light clerical work (e.g. reading or preparing reports).

An operator's work demands sensory perception, selection of signals, long-term and short-term memory, control of data, delivery of conclusions and decisions, and types of manipulation. This type of intellectual work includes the use of research techniques to detect disturbance sources and to make decisions and intervene if necessary. The amount of information must be adapted according to the operator's ability. Visual perception is the most effective human communication channel. Thus, communication between the control system and the operator or dispatcher is established by means of a rationally reduced number of process variables.

Process computers extract facts, rules and knowledge

from process data and help operators in setting the system parameters and putting the system into operation. Therefore the concept and functioning of such expert machines is also presented and evaluated.

4
Generating process data

4.1 PROCESS ENTROPY AND ITS THREE ATTRIBUTES

Process phenomena vary in time, space and use and their versatility is exemplified through various attributes detailing temperature, flow, level, frequency and pressure. Process data characters are created by sensing, measuring and signalling procedures. According to Shannon's communication theory, the mean amount of data generated by each signal character equals the process entropy of

$$H(x) = 1/n \sum_{x_i=x_1}^{x_n} p(x_i) \log_2[1/p(x_i)] \quad \text{bits/character} \tag{4.1}$$

where $p(x_i)$ is the probability of measuring a process attribute equal to x_i, n is the total number of different characters and $\Sigma p(x_i) = 1$. When the probability distribution function $f(x_i)$ of the process attribute value x_i is known or approximated from long-range measurements then the amount of data from equation (4.1) can be given as

$$H(x) = 1/n \sum_{x_i=x_1}^{x_n} f(x_i) \log_2[1/f(x_i)] \quad \text{bits/character} \tag{4.2}$$

The total amount of process data depends on the number of necessary signalling points attached to the process and of the number of events generated in the process. Space and time attributes generate process data almost 'automatically' (Table 4.1), while the usability factor implicates some intricacy.

Example
The entropy of a specific buried pipeline equals $H = 1.369$ bits/character when:

1. The time attribute of the pipe flow is known to be equal to $2\text{m}^3/\text{s}^{-1}$ with the probability $p = 0.9$ and $p = 0.1$ for other cases,

Table 4.1 The entropy of some process attributes

Type	Variable example	Character set	Information rate	Probability distribution	Entropy	Information flow
Time entropy	Temperature transmitter output (mA)	Each 0.1 mA	10 char/s	uniform	7.63 bit/char	76.3 bit/sec
One-dimensional space entropy	Distribution of pressure along the pipe (0–1 bar)	Each 0.1 bar	1 char/m	square root	3.25 bit/char	3.25 bit/m
Two-dimensional space entropy	Distribution of pollutant on Earth's surface (0–10 ppm)	Each 1 ppm	1 char/km^2	exponential	1.5 bit/char	1.5 bit/km^2
Time–space one-dimensional entropy	Time distribution of pressure along the pipe (0–1 bar)	Each 0.1 bar	1 char/ms	square	3.25 bit/char	3.25 bit/ms
Entropy of use	Time distribution of the use of a chemical reactor (on–off)	On–off status signal	1 char/day	$P_{use} = 0.7$ $P_{u\bar{s}e} = 0.3$	0.899 bit/char	0.899 bit/day

Table 4.2 Total entropy calculation of a hypothetic pipeline

Time variable $p('0') = 0.9$ $p('1') = 0.1$	Space variable $p('0') = 0.001$ $p('1') = 0.999$	Usage variable $p('0') = 0.7$ $p('1') = 0.3$	3 $\pi\, p_i$	3 $\pi\, p_i \log (1/\pi p_i)$, bit
t	s	u		
0	0	0	0.00063	0.0067
0	0	1	0.00027	0.0032
0	1	0	0.62937	0.42046
0	1	1	0.26973	0.51
1	0	0	0.00007	0.000966
1	0	1	0.00003	0.000451
1	1	0	0.06993	0.26844
1	1	1	0.02997	0.15169
			3 $\Sigma\, \pi i = 1$	$\Sigma = 1.369$ bit

$H_s = 0.0057$ bit/char $\dfrac{\Delta H_s}{\Delta_s} = 0.9979$ bit

$H_t = 0.235$ bit/char $\dfrac{\Delta H_t}{\Delta_t} = 0.7999$ bit

$H_u = 0.445$ bit/char $\dfrac{\Delta H_u}{\Delta_u} = 0.3999$ bit

    contributing to the time entropy with $H_t = 0.235$ bits/character.

2. The space attribute of the pipe is determined by the pipe being in its predefined position with the probability $p = 0.999$ and $p = 0.001$ for other cases, contributing to the space entropy with $H_s = 0.0057$ bits/character.

3. The usage attribute is determined from the pipe being used with the *a priori* probability of $p = 0.7$ and in other cases $p = 0.3$, adding $H_u = 0.445$ bits/character.

When these process attributes are considered as independent then a composite character set can be created out of these three types giving the total entropy of the independent events as

$$H_T = 1/m \sum_{i=1}^{m} \prod_{j=1}^{3} p_j \log_2 \left(\prod_{j=1}^{3} p_j \right)^{-1} \quad \text{bits/character} \tag{4.3}$$

where m is the total number of event combinations and $\Sigma^m \Pi^3 p_j = 1$ is the example of the total process entropy of the pipe given in Table 4.2. It is obvious from Table 4.2 that there is a minimum of entropy change in the change of the pipe use, which is just the opposite of the result of the separate contribution analysis of the pipe entropy attributes. Thus all independent attributes contributing to the process entropy have to be taken into account when determining the total process entropy.

Generally, when two process variables are dependent their stochastic parts are dependent too, and their entropies are somehow connectable. For the linear case of two process variables v_1 and v_2 when

$$v_2 = a\, v_1 + b \qquad\qquad (4.4)$$

where a and b are known process constants and the distribution function $f(v_1)$ is also known, the distribution function of v_2 equals (Papoulis, 1976).

$$f(v_2) = \frac{1}{|a|} f_{v_1}\!\left(\frac{v_1 - b}{a}\right) \qquad\qquad (4.5)$$

There appear to be two different situations when the process entropy of mutually dependent attributes or events has to be considered. For the case of process usage, being the most complex:

1. Either one or other process unit, part or supplier can be used. This leads to a process conflict situation. The case can also be treated from the standpoint of game theory.
2. Two or more of the process parts, units or suppliers can be used simultaneously. This makes the process situation ripe for optimization by minimizing the process entropy and risk situation, preserving the optimum product and minimizing the product price.

Example
Decide, on the basis of the process entropy, which of the process units should be used when the measured input variable x has the probability density function $f_1 = 0.5$ in the limits $x_1 = 3$, $x_2 = 5$ and the relation

$$v_{2_1} = 2x - 6$$

is valid for the first and

$$v_{2_2} = \frac{3x}{2} - \frac{1}{2}$$

for the second process unit.

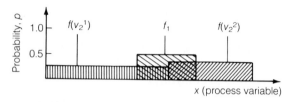

Figure 4.1 The probability density functions of process variables.

The probability distribution functions of the input variable and of both output variables are given in Figure 4.1.

The corresponding entropies are

$$H = f(x) \log_2 \frac{1}{f(x)}$$

where the limits x_1, x_2 are for $f(v_2^1) = (0,4)$ and for $f(v_2^2) = (4,7)$ and

$$f(x) \frac{y - b}{a} = \frac{1}{x_2 - x_1}$$

$$H_1 = \frac{1}{4} \log_2 4 = 0.5$$

$$H_2 = \frac{1}{3} \log_2 3 = 0.528$$

and

$$H_x = 0.5$$

Therefore, by using the first process unit less entropy will be introduced into the process and, under all other equal production aspects, this unit should be preferred.

4.2 DEPENDENT AND INDEPENDENT VARIABLES

Resolving dependency among process variables serves to: (i) reduce redundant measurement equipment; (ii) use redundant data in emergencies; (iii) involve the measure of mutual dependency and (iv) form a basis for process modelling and simulation. Two process variables are independent when there exists an uncorrelated result of their time series data. Regarding practical aspects of the question one has to:

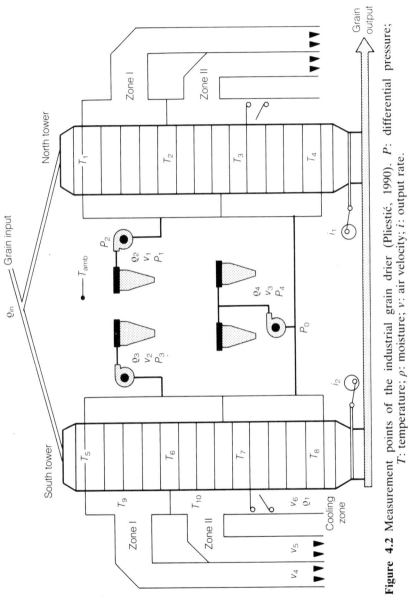

Figure 4.2 Measurement points of the industrial grain drier (Pliestić, 1990). P: differential pressure; T: temperature; ρ: moisture; v: air velocity; i: output rate.

Table 4.3 Measurement of simple data and rank correlation from variables depicted in Figure 4.2* (values are given as a percentage of the total span)

Values				RANKS				RANKS			
T_1	T_5	ρ_{IN}	T_{amb}	R_{T_1}	R_{T_5}	$R_{\rho_{IN}}$	$R_{T_{amb}}$	T_1/T_5	T_1/ρ_{IN}	T_1/T_{amb}	ρ_{in}/T_{amb}
25.35	25.27	31.62	14.29	6.5	6	10	9	0.5	3.5	2.5	1
25.27	25.23	31.84	14.33	10	9.5	4	7	0.5	6	3	3
25.35	25.27	31.67	14.48	6.5	6	8.5	4	0.5	2	2.5	4.5
25.40	25.35	31.70	14.58	3	1.5	7	1	1.5	4	2	6
25.37	25.32	32.06	14.26	4.5	3.5	1	10	1	3.5	5.5	9
25.32	25.23	32.01	14.31	8.5	9.5	2	8	1	6.5	0.5	6
25.32	25.25	31.79	14.55	8.5	8	5.5	2.5	0.5	3	6	3
25.37	25.27	31.67	14.55	4.5	6	8.5	2.5	1.5	4	2	6
25.42	25.32	31.79	14.46	1.5	3.5	5.5	5	2	4	3.5	0.5
25.42	25.35	31.92	14.38	1.5	1.5	3	6	0	1.5	4.5	3
								$D^2 = 11.5$	$D^2 = 166$	$D^2 = 128.5$	$D^2 = 237.5$

$\rho_{T_1-T_5} = 0.930$ $\rho_{T_1-\rho_{IN}} = 0.006$ $\rho_{T_1-T_{amb}} = 0.221$ $\rho_{\rho_{IN}-T_{amb}} = 0.439$

* Samples taken every 4 minutes

1. resolve the dependency within a time longer than the largest process time constant;
2. cope with inherent variable non-linearities;
3. involve the measure of (in)dependency.

The best method to evaluate the dependency of two process variables is to use the Spearman rank correlation method, because it covers the correlation measure and non-linearity between variables (Petz, 1985), the correlation coefficient ρ being

$$\rho = 1 - 6 \times D^2/[n(n^2 - 1)] \qquad (4.7)$$

where D^2 is the squared difference of ranks and n the number of samples taken into calculation.

Example
The measurement set-up of an industrial grain drier is given in Figure 4.2 and measurements of successive samples of four process variables are given in Table 4.3. The correlation coefficients are also given in Table 4.3. As can be seen from the results of the rank correlation, the temperature measurements at the tops of the grain drier are almost fully redundant. Since from the short-term measurement there was no statistical dependency between the moisture measurement and other measurement points, a higher measurement time span was taken into account. According to these data the rank correlation between the moisture measurement and outer temperature was $\rho = -0.439$ indicating that there were definite problems in the installation of the expensive and sensitive moisture measurement instrument. The detection of complete independent variables allows the decoupling of the process into separable sub-processes. The existence of independent variables in processes leads to the effect of additive modelling with complex macro-variables, as shown in Chapters 12 and 13.

5
Variables, states and facts

5.1 VARIABLES AND STATES

Raw process signal data provided with adequate physical parameters produce a process variable. Thus, a process variable is the signal expressed in engineering units. With adequate descriptors in the database it is possible to use process computers in various complex situations, while the process signal changes its value and relation to other process signals. Certain values of crucial process variables can be connected to the process presentation known as the process state.

The working state of a process is usually characterized by its regular production activity. The working state can be detected by the surveillance (supervision) of the values of process variables. The working state can be disturbed by changes in the process hardware, input or output, or by changes in the control hardware or software.

A change in the process state can be influenced and usually limited by technological actions, or by control actions. Technological actions are apparent to each process and their fulfilment can lead the process either into safe or dangerous process states. Safety is measured and estimated as being related to the operator's life, surrounding people and the biological balance of the environment. Technological actions are not influenced by the operator's actions, thus forming a part of a passive protection means.

The possibility of detecting the actual process state by means of detecting process variables, and acting on the process by control actions on process hardware components, serves to change actual process states into new, more preferred process working states. Such control actions can lead the process into allowable or unallowable process states (Figure 5.1).

For some processes the actual process state can be detected by following the values and states of the particular process variables, where two different groups of data can be distinguished according to the type of signal registered (Table 5.1). The analogous signal data

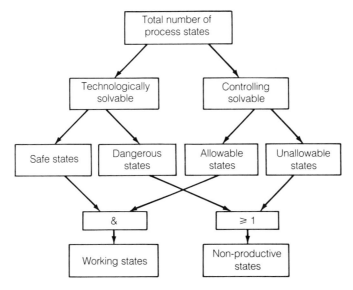

Figure 5.1 The main process states.

for particular data processing function can be used in the sense given in Figure 5.2.

The states of status signals and on/off commands can be combined to obtain a complete picture of safe or dangerous process states (Figure 5.3) and to decide on the action to be executed to obtain a safe state.

Technological actions, the supposed state of process and control, and the operator's actions in the case shown in Figure 5.3 are detailed in Table 5.2. As can be seen, a large amount of preparatory work should be done before safely entering into the automatic control of a very simple basic process unit of the type given in Figure 5.3.

The basic reason for this complexity is that, if given n status process signals of a basic process unit that independently describe independent process variables and process states, then the total of S_T process states equals

$$S_T = 2^n \tag{5.1}$$

All possible S_T states should be analysed and adequate decisions for each of them prepared. Those decisions executed by the operator range from very simple automatic actions (e.g. automatic action on a button after receiving a signal on the video terminal) to sophisticated

Table 5.1 Types of signal registered

Type of signal registered	Content of signal registered	Time/space function relation of process variable
Status signal data	Status information	Not so urgent Not always available Not always up-to-date
	Warning signal	Almost always available data Prompt information
	Alarm signal	Always available Prompt information Usually requires controller action
Analog signal data	Normal to alarm range	Not so urgent data processing Not always available Not always up-to-date
	Alarm to out of range	Data always available Prompt data processing
	Out of range signal	Always available Prompt data processing Usually requires controller action

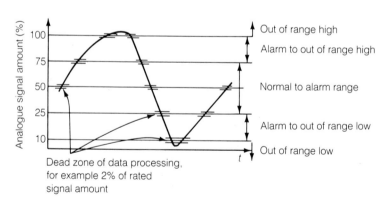

Figure 5.2 An example of the data processing function for analog signals.

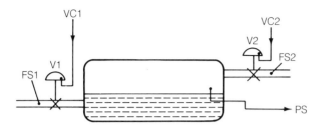

FS1	FS2	PS	supposed process state	actions to be taken
0	0	0	safe state	none
0	0	1	dangerous state	VC2 = '1'
0	1	0	safe state	none
0	1	1	safe state	none
1	0	0	safe state	VC1 = '0'; VC2 = '1'
1	0	1	dangerous state	VC1 = '0'; VC2 = '1'
1	1	0	safe state	none
1	1	1	dangerous state	VC1 = '0'

Figure 5.3 Safe and dangerous process states and corresponding actions to be taken.

FS1 = input flow switch ('1' = FLOW)
FS2 = output flow switch ('1' = FLOW)
PS = pressure switch ('1' = HIGH PRESSURE)
V1 = input valve
V2 = output valve
VC1 = input valve command ('1' = OPEN, '0' = CLOSE)
VC2 = output valve command ('1' = OPEN, '0' = CLOSE)

operator–system dialogues that involve process modelling, emulation and real time expert system consultancy.

Even more process states than are given in equation (5.1) appear when analogous and counter state signals are taken into account. Thus, a recognition of process states on the side of the operator or expert system software should be preferred since a large amount of process states is expected.

Distinguishing k groups of signals, each group with an equal number of I_i states, the total amount of states S_T equals

$$S_T = \prod_{i=1}^{k} I_i^{N_i} \tag{5.2}$$

where N_i is the number of signals in the ith group (each with I_i states). A process unit with two analogous signals each with five states and four binary signals possesses $5^2 \times 2^4 = 2000$ different states.

Table 5.2 Control of a simple basic process unit

Actions	Process state number	Status signals FS1	FS2	PS	Control and operator action	Supposed process state
None	1	0	0	0	None.	Tank is empty
Pressure on tank above safety level. Upon opening of valve V_2 the tank is emptied, which should be checked through flow switch FS	2	0	0	1	Open valve 2 in order to decrease pressure. If after given time interval T_2 the pressure does not decrease, check pressure switch.	Tank is full above safety level
Emptying of the tank is taking place	3	0	1	0	No control action. If flow switch state equals zero then the tank is supposed to be empty.	Tank is being emptied and the pressure low
Emptying of the tank is taking place	4	0	1	1	None. If state 4 lasts more than given time interval T_2 then check pressure switch and flow switch.	Tank is being emptied and the pressure high

No.	Condition				Action	Comment
5	Tank is being filled	1	0	0	If tank is being filled for more than time interval T_2 and pressure is still low then close valve V_1, open valve V_2 and check pressure switch PS.	Tank is being filled above safety level
6	Tank is being filled	1	0	1	Close valve V_1 and open valve V_2 in order to decrease pressure. If after time interval T_3 the pressure is still high then check pressure switch PS and flow switch FS2.	Tank is being filled above safety level
7	Tank is being filled and emptied at the same time	1	1	0	None.	Tank is being filled and emptied at the same time below safety limit
8	Tank is being filled and emptied at the same time	1	1	1	Close valve V_1 in order to decrease pressure. If after time interval T_4 the pressure is still high then check the pressure switch PS and the state of valve V_1	Tank is being filled and emptied at the same time above safety limit

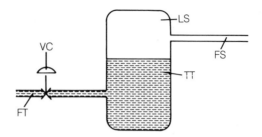

Figure 5.4 A simple hypothetical basic process unit.
VC = valve command
FT = flow transmitter
TT = temperature transmitter
FS = flow switch
LS = level switch

The main point in direct communication between the operator and the process is to allow a stratified approach to the data presentation and print-out. The most important fact in fast decision making by the operator is to provide exact information about the useful facts about the process state. Thus, if the operator is provided with a better protocol for the presentation of the useful facts about the process state, the overall process control exhibits a higher information efficiency I_e that can be presented as

$$I_e = S_T/S_u \qquad (5.3)$$

where S_T is the total number of possible different process states to be presented to the operator that completely describe the process, and S_u is the number of different process states actually presented to the operator and differing from the normal state in one element; the operator then decides what information is useful about the process state in order to decide about the presented abnormal state. An example of the calculation of I_e for a particular simple, hypothetical basic process unit can be based on the data given in Figure 5.4. The total number of possible process states given in Table 5.3 equals the product of all process variable states. Let us suppose that the presentation of a change of state is sufficient for the decision on the actual process state. Since seven possible changes of state events can be expected, supposing the process is in the normal state concerning the analogue signal values, the information efficiency I_e in this example equals $72/7 \approx 10$. The transitions to alarm process states can be supported by the time tag, so that the obtained print-

Table 5.3 Total number of process states

Process variable	Number of normal process states	Number of alarm states	Total number of variable states	Total number of possible new states (deviation from normal state)
Valve command	2	–	2	1
Flow transmitter	1	(high) 2 (low)	3	(normal to high) 2 (normal to low)
Temperature transmitter	1	(high) 2 low	3	(normal to high) 2 (normal to low)
Flow switch	2	–	2	1
Level switch	2	–	2	1

Total number of process states: $S_T = 2 \times 3 \times 3 \times 2 \times 2 = 72$
Total number of possible new states differing from normal states: $S_U = 7$

out can be used for official supervision and arbitration. Fast, complex and mutually interconnected processes (e.g. electrical power generation, transmission and distribution systems) are then provided with such functional recording systems, usually termed chronological event recorders, that are primarily used for the time detection and time-of-event appearance discrimination of relevant process data and the operator's actions. Such chronological event recorders are usually installed for a particular process control, or even for a basic process unit. For large interconnected systems, exact time distributions have to be provided to satisfy the demand for precise time-of-event control in a process.

5.2 FABRICATING AND USING FACTS

Fact is a data triplet composed of an object, attribute and value, the object being a physical or conceptual entity and the attribute a characteristic feature or property. Process signals, variables and states serve to present actual process facts that are otherwise unobservable. The extraction of facts is used for proper process control and surveillance because by using facts and only facts the operator's

tank–1, a water-tank	
Notes	OK
Names	TANK–1
Volume	67.987
Level	70.0
Area	75
Level status	OK
Low level line	25 feet
High level line	100 feet

Figure 5.5 An example of fixed facts and its content.

knowledge can be synthesized for decision making. In addition, the automation of process decision making can be done on the knowledge bases in principally the same way. The simplest definition states that a fact is an observable phenomenon in a process. Some process facts are easy to obtain but some are very intriguing.

Example
In the case of a buried pipeline the state of corrosion is of prime interest for the budget, economic analysis and spare parts storage. The essence of the problem is the prediction of the state of pipe corrosion, the so called CSI (Corrosion Status Index) (Kumar, 1988) which is given with the expression

$$CSI = 100 - 70 \,(age/life) \, \exp(1.724) \qquad (5.4)$$

where

$$life = CTR^{0.05}\exp(0.13\,pH - 0.41\,M - 0.2265\,S) \qquad (5.6)$$

where C is a constant, T is the thickness of the pipe wall, R is the electric resistance of the soil, pH is the pH value of the soil, M is 1 for saturated soil and 0 for other soils and S is the sulphide and moisture content. The predicted year of the first leak is calculated using the data and formula (5.4), which is the only fact of interest to the owner of the pipeline and which can otherwise be obtained only by expensive and cumbersome measurements at the site.

Fixed facts about process equipment are typed in for the expert system's use in the preparatory phase of the system design. An example of fixed facts and their content is given in Figure 5.5. Variable facts depend on process variable states and their appearance has to be modelled in order to be detected and connected to the hypotheses or decision trees. The models needed for their calculation have also to be given to the expert system. The connection of models for fact evaluation and confidence in the semantic value of the given process information is given in Chapter 14.

6
Process presentation

6.1 MEANS AND EFFECTS OF PRESENTATION

The visualization of all pertinent process control data is performed by alphanumeric, semi-graphic and graphic display units, and purpose-built operator consoles, panels and mimics.

A survey of audio response, semi-graphic and graphic display units is outlined in Table 6.1.

The print-out of all pertinent process control data and the logging of data are performed by the printer, line printer or other hard copy unit.

Thus the visualization and data logging can be broken down into two areas:

1. hardware equipment for the visualization and data logging;
2. protocols for the process data visualization and data logging.

Since the same protocols can be applied to different equipment, the main issues in the visualization and data logging are protocols for man–machine communication. These protocols are implemented differently for various processes, since the main task of the process data visualization and data logging is to allow the best direct communication of the operator with the process and the fastest possible decision making by the operator.

The most efficient data visualizations are obtained by using semi-graphic and graphic video terminals where process mimic diagrams can be shown and provided with up-to-date process data. Such an interface allows the presentation of the process to the operators in a way in which they can:

1. see the actual structure of the process and the interactions between specific basic process units, mainly by means of the static part of the mimic diagram;

Table 6.1 The main features of audio response, semi-graphic and graphic display units

Features	Audio response unit	Semi-graphic display unit	Graphic display unit
Assisted computer system (ACS) or controller	Microcomputer	Microcomputer ≤64 k bytes refresh	Graphic processor ≥2–4 k words
Connection to ACS or controller	Parallel interface Serial interface I/O bus connection	Integrated into the system serial interface RS232C (HDX, FDX) 300–9600 bd	Integrated into the system
Minimum– maximum number of characters, points of words expressed	10–10000 words unlimited <300 phrases	64 characters 64 special generated symbols 20–140 lines per unit 60–80 char/line graphic matrix 1024×512 dots 512×256 dots	95 ASCII + 27 symbols
Selection of expression	Words/phrases selectable	4–64 intensity levels 7–9 colours foreground (7–9 colours background)	16–24 intensity levels 7–9 colours 240 character sizes
Expression timing	Word length 0.5 s phrase length 1.6 s	60 refresh/s	$5\,\mu s$/character <40 km/s vector
Expression generation technique	Digital compressed analogue, recorded on film strips, speech synthesis	Dot matrix	Vector drawing
Options and console	Selection of words selection male/ female any language	Zoom Selective erase Reverse video typewriter Cursor pad automatic plotting	Typewriter, cursor pad 128:1 zoom Depth cueing ≤128 additional user defined characters Hardware generated circles and arcs console devices like control dials, joystick, digital tablet, additional refresh buffer memory

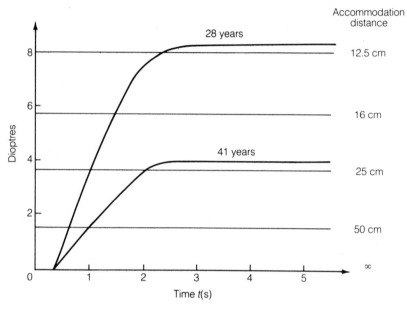

Figure 6.1 The accommodation step response with time and age of operator for yellow–green colour (Krueger and Muller-Limmroth, 1981).

2. estimate the quantitative relations between process variables mainly by means of a variable part of the mimic diagram that presents actual and historic process model data.

The ability of a mimic diagram to allow fast magnification of process details by calling mimic diagrams of specific process parts of complex processes increases the speed of decision making by the operator in the case of the vast numbers of process data. The efficiency of visual presentation can be increased by approximately 40 per cent in information content (Z. Smrkić, personal communication) by a colour video display unit. But the human eye has different reaction times, depending on colour. The cause possibly lies in chromatic aberration where corrections of about +2 dioptres are needed for red and −2 for blue (Figure 6.1). Thus, yellow–green video display screens are preferred for processes where fast reactions are expected from the operators.

Colours possess a strong influence on human behaviour and therefore colour video displays are preferred for many process control applications when appropriate colours are used. Some common colour codes and their meanings are given in Table 6.2. The use of colour for actual process control purposes is given in Table 6.3.

6.2 THE COMMUNICATION OF STRUCTURE: PROCESS FORM

Process presentation is based on the process symbolic model as a part of the presentation scheme, Figure 6.2, allowing the communication of even the most complex object and relations in a very natural way (Grassi, 1957; Hinde). The symbolic model contains a synthetic idea of the process structure with a meaning which is obvious to the user. However, the quality of the symbolic model mostly depends on the designer's understanding of the process structure (Platon, 1988). Two problems still remain:

1. the actual process form is not available to the operator since the process objects are distant and scaled disproportionately and the processes are invisible and sometimes incomparable (i.e. the subject of internal non-comparison);
2. the process inside the object represents a hierarchy of relations and its form should be dynamized in order to be more familiar to the operator.

Thus many different aspects of process presentation should be mastered: pre-processing, frame, dynamics, editing, sound participation, process script and theory of presentation (Gianetti, 1987).

Pre-processing mostly uses filtering, scaling, windows and fast data acquisition in order to compress data, extract the signal, suppress noise, cancel drift (Äström, 1984).

The process frame includes all the space and set-up in the control room where process data and commands are presented, such as blind schemes, control panels, control table, mimic diagrams and print-outs.

The inter-frame and in-frame designs comprise:

1. dominant process contrast for immediate attraction of the operator's attention (due to conspicuous and compelling contrast) such as a sudden change of colour on the control panel buttons;
2. subsidiary contrast that next attracts the operator in a specific sequence deliberately programmed by the frame structure, such

Table 6.2 Some common colour codes and their meanings

Features	Colour							
	Red	Yellow	Green	Blue	Purple	White	Grey	Black
Positive associations	Life, warmth, passion, valour, sentiment	Sun, light, intuition, intellect	Vegetation, nature, sympathy, prosperity, hope	Sky, day, sea, thinking, devotion, truth	Power, spirituality, royalty, empire	Day, innocence, purity perfection	Maturity, discretion humility	Mighty, stark dignified, night
Negative associations	Spilled blood, burning, wounds, war	Treachery, cowardice	Death, lividness, envy, disgace	Dark blue denotes night and stormy sea, doubt	Sublimation, martyrdom, regret	Spectral, ghostly, cold, void	Neutralism, egoism, depression, indifference	Morbidity, despair, night, evil, sin
Most common cultural meaning	Colour of joy and festive occasions. Fighting, anger, danger. Marriage colour (folklore). Protection from death (folklore). Fever or protection from disease (folklore). Holidays	Quarantine colour (medicine). Emperor colour (China). Happiness (Egypt). Gay (Japan). Sensationalism (journalism).	Feminine (American Indians). Fertility, vegetation (Egypt). Youth, energy (Japan). Jealousy, envy (language). Sterile (medicine).	Virtue, faith, truth (Egypt). Ghost, fiend (Japanese theatre).	Worn by figures (China). Virtue and faith (Egypt). Rage (language). Winner (prizes).	Virtue and purity (fashion). Death (with black). White flag (surrender, peace).	Old, mature (psychology). Wisdom (Judaism).	Death, winter (China). Water (China). Black cat (superstition). Evil, despair (language). Morbid (psychology).

Most common technical meaning	Flammable gas or liquid. Stop (traffic). Plus of power supply (electrical engineering). Roads, telegraph lines (geography).	Transistor emitters (electrical engineering). Heating circuits Oxidizing agent. Radioactive matter. Caution (motor racing). Warning of danger (safety). General warning. Gas (piping).	Control grids wiring chassis (electrical engineering). Nitrogen, compressed gas. Clear or go signals, permission (traffic and safety). Oil (piping).	Transistor collectors (electrical engineering). Caution on repair (safety). Protective materials (piping). Sea (geography).	Power supply minus (electrical engineering). Radiation hazards. Valuable materials (safety).	Bias supply wiring chassis (electrical engineering). Regulation (traffic).	AC power lines (electrical engineering). Steam (piping).	Grounds on wiring chassis (electrical engineering). Contours (geography). Corrosive material (handling of goods).
Colour preference	Second (adult) Fourth (child)	Eighth (adult) First (child)	Third (adult) Seventh (child)	First (adult) Sixth (child)	Sixth (adult) Eighth (child)	Fourth (adult) Second (child)	—	—

Table 6.3(*a*) The use of colour in process control with neutral background

Data priority	Process status	Colour
Highest priority	The part of process chosen for operator control	White
	Refreshment of data blocked	Violet
	Unaccepted alarm	Red flashing
	Equipment failure	Red
	Process control function blocked	Yellow function
Lowest priority	Normal process state	Green or dark blue

Table 6.3(*b*) The use of colour in process control with background effect

Data priority	Process status	Colours
Highest priority	Out of alarm range	White on red
	Out of disturbance range	Black on yellow
	Out of tolerance range	Yellow on blue
	Disturbance in analogue control	White on red
	Disturbance in on–off control	Black on yellow
	Measurement disturbance	Yellow on blue
Lowest priority	Process signalling	White on red

as a change in the mimic content, by following the change in the control panel buttons, then the event recorder print-out etc.;
3. hierarchy of space giving dominance to central places in the control room and the central space on the screen;
4. proximity pattern, where an object presented at a smaller size tends to be more objectively accepted by the operator;
5. open versus closed form (Figure 6.3); open form suggests a more realistic process display, while today a more often encountered closed form tends to emphasize an unfamiliar and formal approach.

Process presentation dynamics follows important process changes in time and place. Special effects of presentation dynamics can be produced by the distortion of dynamics, such as a jump and freeze-

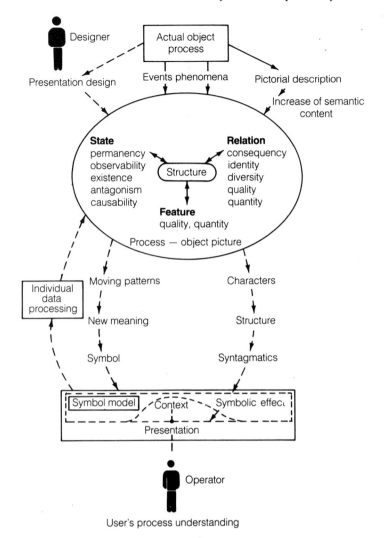

Figure 6.2 The presentation scheme.

frame. Direction of change on the frame is also psychologically important since a movement on the frame from left to right indicates a favourable context of the process change.

Editing ensures:

1. continuity by presenting process sequences as they actually happen in time;
2. parallelism by presenting two or more concurrent process actions

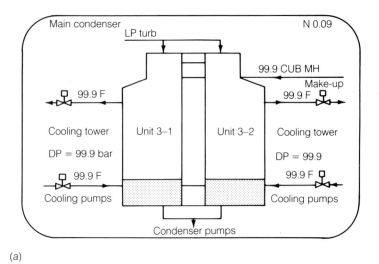

(a)

(b)

Figure 6.3 (*a*) A CRT display of a process state – closed form of a mimic diagram; (*b*) a CRT display of a process state – open form of a mimic diagram.

using windows, a split screen, many screens or similar technical aids;

3. flashback by historic data reproduction, such as the *post mortem* review of a fatal accident;
4. dialecticity by contrasting and opposing process facts which are in competition relating to process resources, such as a fuel–air burner towards feed water–steam–power cycles in power plants;
5. objectivity, by bringing all relevant data and trends to the operator's room for selection.

Sounds from the process sites are rarely used, although a few types may be very important, such as the burner noise in a power plant; speech communication is usually over-used, just like alarm sounds, which can be very intrusively designed. Music is used to relax the crew and to improve the usual unrealistic effect of the control place.

The process presentation script is a simple description of what should be brought to the operator at each stage of process development, leaving the dramatic context to the trained operator. The description contains goals, dialogues, means and procedures available for the required process action.

Process presentation theory prefers an eclectic, semantic and synthetic approach, forcing semantically valuable process events to be emphasized in the process design.

In order to make a fast selection of process data possible the presentation is usually organized between two and four hierarchical levels.

First level: The operator is informed that the system is working without perturbation and what has to be searched for with the highest priority in case of perturbation.

Second level: The structure of the process or plant that is controlled is exhibited, indicating the principal energy and material flow and processing states and including an alarm summary with the last received alarm.

Third level: The operator is allowed to see those process details that actually exist and those process and alarm states that have actually happened.

Fourth level: The operator is provided with additional data for the supervision and control of specific process components, e.g. analogue values, limit values and zone values of a specific process variable.

Table 6.4 details a typical four-level process data presentation for normal and alarm process conditions.

Table 6.4 Four-level data presentation for normal and alarm conditions

Information type	Normal process conditions			Alarm conditions		
	Information level	Type of display	Display hardware	Information level	Type of display	Display hardware
Status of process connection and configuration state:		Permanent display		1	Alarm is always spontaneously indicated in the form of acoustic signal, general alarm indication, flashing light of the object on a CRT screen	Acoustic indicator, gong, CRT display, Operator's table
hardware connecting elements (position of valves, breakers etc.)	1	Elementary	Operator's table CRT display			
bulk process hardware elements (transformer sections, manifolds, vessels etc.)	2	Compressed	CRT display			

indicators (state sensors, switches, etc.)	3	Complete on call	CRT display	3	Clear text on alarm field of CRT screen	CRT display
Analogue values: process primary values such as voltage, current, flow, level etc.)	1 2 3	Not displayed		1	Alarm indication when passing of limit value	CRT display Operator's console
process global values such as net frequency, valve position, etc.	4	Completely displayed. Permanently displayed. Possible selection.	CRT display. Measuring instrument. Analog recorder. Numeric indicator. XY plotter.	3	Text indicating the cause of perturbation on CRT alarm field	CRT display
Counter state, Increments	4	Possible selection	Measuring instruments. Counter display.		Usually not displayed	

Table 6.5 Types of report

Information type	Content	Availability
1. Events		
Exploitation events	All operator actions such as commands, alarm quittings, all process spontaneous changes, signalling alarms, passage through limit values	Spontaneous at the moment of appearance
Alarm events	All alarm conditions.	On demand
Process configuration	Overview of all process connection in a real configuration.	On demand
Event recording	Print-out of perturbations in chronological order	Automatic when provided with event recording. On demand when event recording in a mass memory.
2. Measuring values		
Communication of measuring values.	Issuing of measuring and limit values.	Periodically or on demand.
Supervision of measuring values.	Issuing of selected measuring values.	Periodically or on demand.
Statistics.	Recording of mean and maximum values.	On demand and often graphically on CRT screen or on analog recorders.
3. Counter state.		
State of process variable counters	All counter states	Periodically or on demand

An important view of the operator–system communication is established by the report issuance that provides the process state documentation and the initiation of necessary process state calculations. Report issuance can be divided into the process event documentation and collective process data issuance (Table 6.5).

REFERENCES FOR PART TWO

Äström, K.J. (1984) *Computer Controlled Systems*, Prentice Hall International Inc. Englewood Cliffs, N.J.

Gianetti, L.D. (1987) *Understanding movies*. Prentice-Hall, Englewood Cliffs, N.J.

Grassi, E. (1957), *Kunst und Mythos*. Hamburg.

Hinde, R.A. Model and concept of 'drive', *Br. J. Phil. Sci.* **VI**, 24, p. 323.

Itten, J. (1973) *The art of color*. Van Nostrand Reinhold, New York.

Krueger, H. and Müller-Limmroth, W. (1981) *Arbeiten mit dem Bildschirm-aber richtig*. Bayersche Staatsministerium für Arbeit und Sozialordnung, München.

Kumar, A. (1988) *Pipeline industry*. October, 19–23.

Macrae, N. (1988) *The Economist-The World in 1989*. p. 88.

Papoulis, A. (1965) *Probability, random variables and stochastic processes*. McGraw-Hill, New York.

Petz, B. (1985) *Basic statistical methods for non-mathematicians*. SNLiber, Zagreb (In Croatian).

Platon, (1988) *On language and cognition*. Rad, Belgrade (In Serbo-Croat, 2nd edn).

Pliestić, S. and Dobriče vić, J. (1991) *Computer control of gravitational driers*, Proc. of the vii Int. Meeting of Drying and Storing Technology, Tuheljske Toplice, Faculty of Agricultural Sciences, Zagreb, 33–44.

Woodson, W.E. and Conover, D.W. (1966) *Human engineering guide for equipment designers*. University of California Press, Los Angeles.

Part Three
Dialogue

Dialogue is a trial and error procedure between correspondents to solve a common problem. An exchange of unknown data and relations can be performed by a dialogue. Even structures can be communicated in this manner, using common language elements and pictures as structure representatives. Problem solving is the feature both of people and machines. It is based on the problem solving search of the knowledge base by applying inference mechanisms and rules and by coding and changing internal states. The important point in the inference speed is the entropy concept.

Procedures are a very important part of the dialogue, since they include some common ways of internal coding, communicating and presenting data and structures, and above all the manner and methods of exchanging data and structures in a dialogue process.

Dialogues and procedures are performed using complex data structures and the required decisions should be justified in the expert system environment.

The sideways chaining procedure in binary expert systems is exemplified.

7
Facts, rules and decisions

7.1 FACTS AND RULES – THE BINARY CASE

Inference is an intriguing task. Sometimes it is connected with exact relations of the type IF–THEN–ELSE, for example in programmable logic controllers implemented for process control. A more complex and less exact kind of inference is an attribute of human-like inference mechanisms. Logic judgements are used for events that are stochastically unconnected. Inexact inference is used for mutually connected appearances – facts. While for exact inference one can easily follow the procedure, we can easily understand only the manner of inexact inference.

Most technical inexact inference mechanisms are based on the Bayes' statistics and entropy concept (Brillouin, 1956). Bayes' theorem states that even for the most improbable event there is a prior probability of occurrence and the fact of a detected event is connected with the posterior probability in the following way

$$P(H) + P(E) \Rightarrow P(H:E) \tag{7.1}$$

where $P(H)$ is the prior probability, $P(E)$ is the probability of occurrence of the event in the concrete case and $P(H:E)$ the posterior probability of the event's occurrence. By using the relations

$$P(H:E) = P(E:H)\, P(H)/P(E) \tag{7.2}$$

$$P(E) = P(E:H)\, P(H) + P(E:\text{not } H)\, P(\text{not } H)$$

where $P(E:H)$ written as p_y, means the probability of the event occurring in the case of the hypothesis being true, and $P(E:\text{not } H)$, written as p_n, means the probability of the event occurring in the case of the hypothesis not being true. By using the relations (7.2), we can obtain the formats used in the knowledge base

$$P(H:E) = p_y\, p/[p_y\, p + p_n\, (1 - p)] \tag{7.3}$$

$$P(H:\text{not } E) = (1 - p_y)\, p/[(1 - p_y)\, p + (1 - p_n)(1 - p)]$$

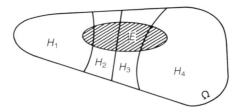

Figure 7.1 Hypotheses and events.

where p stands for $p(H)$.

Having calculated a new $P(H:E)$ from a newly acquired fact, one can forget the original $P(H)$ and instead use this new $P(H:E)$ as a new $P(H)$. The same applies also to $P(\text{not } H)$. Thus, the whole process can be repeated time and again, a different probability being used as derived from the last posterior probability.

Hypotheses form a complete mutually exclusive family when

$$\bigcup_{i+1}^{n} H_i = \Omega$$

($\Omega =$ universal set) and

$$\sum_{i+1}^{n} P(H_i) = 1$$

For E being any event the following relation is valid (Figure 7.1),

$$(E \cap H_1) \cup (E \cap H_2) \cup (E \cap H_3) \cup (E \cap H_4) = E \qquad (7.4)$$

Supposing that events $E \cap H_i$ ($i = 1, \ldots, n$) are mutually exclusive. This gives

$$\sum_{i=1}^{n} P(E \cap H_i) = P(E) \qquad (7.5)$$

Applying the relation

$$P(E \cap H_i) = P(E) \, P(H_i:E) = P(H_i) \, P(E:H_i)$$

we obtain

$$\sum_{i=1}^{n} P(H_i) \, P(E:H_i) = P(E) \qquad (7.6)$$

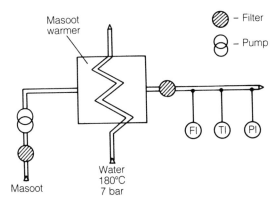

Figure 7.2 The functional scheme of a masoot warmer in a power plant.

Table 7.1 Knowledge data base of the binary expert system

Hypotheses $p(H_1)$		Event facts			
		Masoot temperature high	Masoot temperature low	Flow OK	Pressure OK
Failure of the massot	Yes	0.2	0.8	0.5	0.4
warmer $p_{H_1} = 0.05$	No	0.9	0.1	0.9	0.5
Failure of hot water	Yes	0.05	1	–	–
supply $p_{H_2} = 0.01$	No	0.999	0.05	–	–
Masoot filter failure	Yes	0.4	–	0.7	–
$p_{H_3} = 0.2$	No	0.2	–	1	–
Masoot pump failure	Yes	0.8	–	0	0
$p_{H_4} = 0.01$	No	0.01	–	0.99	0.95

$$P(H_4:\bar{E}) \quad P(H_4:E)$$

The probability of event E can be calculated if we know the probability of H_i (usually called the hypothesis) and the conditional probabilities of events E in relation to the hypothesis H_i ($i = 1, \ldots, n$). By using relations (7.2) and (7.6) we obtain

$$P(H_i:E) = P(H_i)\,P(E:H_i) \left(\sum_{j=1}^{n} P(H_i)\,P(E:H_j) \right)^{-1} \qquad (7.7)$$

which is the Bayes formula applicable to different practical tasks in which the probability of the conditional hypothesis H_i is searched for when the event E has happened which usually appears with one of the hypotheses H_j ($j = 1, \ldots, n$).

Example

The case of a masoot warmer in a thermoelectric power plant is shown in Figure 7.2, where four different events are taken into account and put into the probability relation with four hypotheses. This is a case of the binary expert diagnostic system. The corresponding data in the knowledge base are also given in Table 7.1.

In order to apply formula (7.7) the hypotheses have to make a complete family of events, i.e. no event can happen outside the hypotheses, taking into account all known facts.

7.2 NUMBER OF RULES AND QUESTIONS AND EVALUATION CRITERIA

The calculation of each hypothesis connected to a certain fact gives a change in the prior probabilities, according to formula (7.2). There appear two problems for expert systems in process control:

1.   the number of rules connecting facts and hypotheses;
2.   evaluation criteria.

For the binary expert system, if there are f facts, then the maximum number of hypotheses H_M is given as

$$H_M = f(f - 1) + 1 \tag{7.8}$$

The maximum number of checkings (questions) Q_M of all hypotheses equals (also for binary expert systems)

$$Q_M = f \left(\sum_{i+1}^{f-1} i + 1 \right) \tag{7.9}$$

For larger process units one has to collect logical implications, hypotheses, ckeckings, process facts, rules and facts in a complex structure, given in Figure 7.3. The probability conclusion needs the elaboration of evaluation criteria.

A simple evaluation criterion is given as the rule value (Kononenko *et al.*, 1986)

$$RV = \sum_{i=1}^{n} |P(H_i : E) - P(H_i : \text{not } E)| \tag{7.10}$$

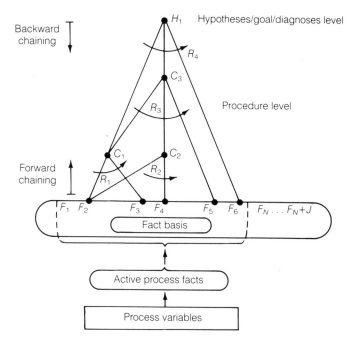

Figure 7.3 Facts, active facts, hypotheses and rules.
C: probabilistic conclusion
R: Bayes' rule
H: hypothesis

By calculating the rule value with formula (7.10) from the example given in Table 7.1, the data given in Table 7.2 can be obtained. This calculation gives an indication of the most probable hypothesis. By scanning the facts for activity or by asking about their state, it can be found that FACT1 = '0', so that the new prior Bayes probabilities can be calculated as given in Table 7.2, new rule values obtained etc.

The activation of each fact drastically changes probabilities and evaluation results. Therefore, when two or more facts are activated the rule with more connections to these facts will in practice, give the most dominant evaluation. The sequence of search should be pointed towards the item of evidence (fact) which can induce the greatest probability shift in all hypotheses under consideration. Questions stored in the knowledge base should be responded to either by the operator or automatically by the system, for example using different models for specific process situations (ABB, 1989),

Table 7.2 The calculation of a rule value (side ways chaining) for the case of the masoot warmer in Figure 7.2

$RV_{FACT1} = 2.33501$	Masoot temperature high
$RV_{FACT2} = 0.45279$	Masoot temperature low
$RV_{FACT3} = 1.52778$	Flow OK
$RV_{FACT4} = 0.18703$	Pressure OK

Question: Masoot temperature high?
If answer no, calculation of new probabilities

$p_{H1NEW} = p(H_1 : \bar{E}) = 0.29629$
$p_{H2NEW} = p(H_2 : \bar{E}) = 0.90562$
$p_{H3NEW} = p(H_3 : \bar{E}) = 0.15784$
$p_{H4NEW} = p(H_4 : \bar{E}) = 0.00204$

And there follow new values of the rule values:

$RV_{FACT1} = 0$	Masoot temperature high
$RV_{FACT2} = 1.67363$	Masoot temperature low
$RV_{FACT3} = 1.11379$	Flow OK
$RV_{FACT4} = 0.12283$	Pressure OK

Thus one has to ask for the fact of low temperature of masoot, etc.

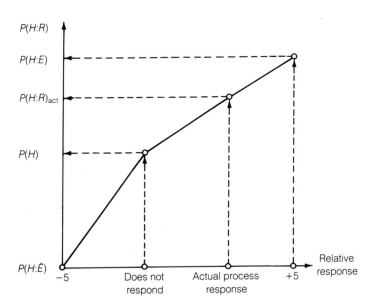

Figure 7.4 Uncertainty in process response.

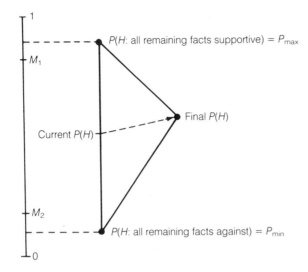

Figure 7.5 The five probabilities used for a given hypothesis.

and involving the finer measurement tunings and control procedures in order to find the most probable hypothesis. The response variable R can be scaled, for instance, between -5 and $+5$ and all hypotheses referred to that item of evidence in their knowledge base have to be corrected to find $P(H:R)$ as (Figure 7.4)

$$P(H:R) = P(H:E)\ P(E:R) + P(H:\text{not } E)\ P(\text{not } E:R) \qquad (7.11)$$

The rule values for all items of evidence should be recalculated to allow the change in the probabilities that have taken place given the last process response. Then the minimum and maximum values which each hypothesis may still attain should be calculated (Figure 7.5). Essentially there are five quantities for each hypothesis; each of these quantities is a probability, i.e.

$P(H)$ is the current estimated probability of that hypothesis being true;

$P(\text{max})$ is the current maximum probability that this particular hypothesis could attain if all the remaining facts went in its favour;

P(min)   is the current minimum probability which a particular hypothesis could attain if all the remaining facts worked against it;

*M*1   is the upper threshold criterion for accepting a particular hypothesis calculated as a proportion of *P*(max) before any fact has been acquired at all.

*M*2   is the lower threshold criterion for rejecting a particular hypothesis calculated as a proportion either of *P*(max) or *P*(min) before any fact has been acquired at all.

The most likely outcome is found if there is some hypothesis for which *P*(min) is greater than *P*(max) for any other hypothesis. The likely conclusions are those hypotheses for which *P*(min) is greater than *M*1.

Uncertain conclusions are those hypotheses for which *P*(min) is lower than *M*1 and *P*(max) is greater than *M*2. These items have potential for resolving the uncertainty.

No inference is possible (or the hypotheses are false) for those hypotheses where *P*(max) is lower than *M*2.

The flow chart of the procedure proposed is given in Figure 7.6.

7.3   ORDERING PROBLEMS AND CLASSIFICATION ENTROPY

There are *f*! (*f* factorial) different ways of ordering the question asking process for *f* facts. By applying the entropy concept to the ordering of questions we can get a much more efficient decision procedure. It can be done by measuring the uncertainty of the classification according to a different ordering of facts. As the entropy increases, the amount of information gained by the knowledge of the final classification increases. What we are really interested in are the results of the plant behaviour. The behaviour can usually be classified into n classes, C_1, C_2, ... C_n, (Table 7.3). Thus, the entropy of classification, $H(C)$ is

$$H(C) = -\sum_{i=1}^{n} p(C_i) \log_2 p(C_i) \tag{7.12}$$

where $p(C_i)$ is the probability of the plant being in a class result C_i. The uncertainty about the efficiency of the power plant going up or down, based on the data in Table 7.3(*a*), equals $H(C) = 1$ since $p(\text{efficiency} = \text{up}) = p(\text{efficiency} = \text{down}) = 0.5$.

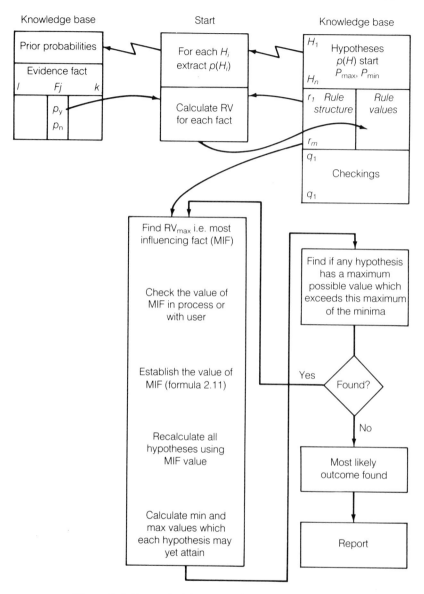

Figure 7.6 Bayes decision procedure and knowledge base.

The calculation of the entropy of classification of an attribute of the fact is done according to Table 7.3(*b*) for fuel by the expression.

$$H(C_i/a_j) = -\sum_{i=1}^{n} p(C_i/a_j) \log_2 p(C_i/a_j) \tag{7.13}$$

Table 7.3(*a*) Table of facts for the example of a power plant

Result	Fact		
Efficiency (η)	Age	Fuel	Type of plant
Down	Old	Coal	Mixed
Down	Mid-life	Gas	Mixed
Up	Mid-life	Coal	Electric
Down	Old	Coal	Electric
Up	New	Coal	Electric
Up	New	Coal	Mixed
Up	Mid-life	Coal	Mixed
Up	New	Gas	Mixed
Down	Mid-life	Gas	Electric
Down	Old	Gas	Mixed

Table 7.3(*b*) Sub-table of Table 7.3(*a*) for the attribute 'Fuel'

Fuel	Efficiency
Coal	Down
Coal	Up
Coal	Down
Coal	Up
Coal	Up
Coal	Up
Gas	Down
Gas	Up
Gas	Down
Gas	Down

where n is again the number of result classes and function $p(C_i/a_j)$ is the probability that the class value is C_i when the attribute a has the jth value.

Actual entropies for Table 7.3(*b*) are

$$H(\text{efficiency/fuel} = \text{coal}) = 0.918 \qquad (7.14)$$

$$H(\text{efficiency/fuel} = \text{gas}) = 0.811$$

The total entropy of attribute A equals

$$H(\text{result}/A) = -\sum_{j=1}^{m} p(a_j)\, H(A/a_j) \qquad (7.15)$$

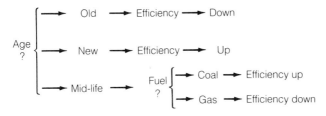

Figure 7.7 The decision tree for the example of power plant efficiency.

where *m* is the total number of values for attribute *A*.

Thus the entropy of classification after choosing a particular attribute *A* (fuel) is the weighted average of the entropy for each value of the attribute:

$$H(\text{efficiency/fuel}) = (6/10)\,0.918 + (4/10)\,0.811 = 0.8752$$

Applying the procedure to two other attributes, the following amounts can be found: $H(\text{Efficiency/Age}) = 0.4$ and $H(\text{Efficiency/Type}) = 1.0$. Since $H(\text{Efficiency/Age})$ gives the smallest entropy and the least uncertainty, age is the best attribute to select for the initial split of the decision tree. The result of the decision tree for the given example is shown in Figure 7.7. Practical application of the decision tree using the entropy concept is given in the decision tree in Figure 13.12.

8

Dialogue – the operator's standpoint

A dialogue in process control is performed at least from the operator's side requiring process data. The process computer usually supplies these data. A dialogue is needed to check the operator's behaviour on the part of the system as well. There are internal states in both the process and the process control computer. These internal states are exhibited through their external states which are sometimes coded variables and sometimes very simple external variables. Both internal and external states are mutually connected and in permanent change.

Operators also possess internal subconscious states and external levels and states which are changing as well. What matters in process control is the operator's knowledge of the system states and the actions to be taken.

Thus, the problem here is one of data reduction, and it is useful to consider the following systems on the basis of their size, since this affects the amount of information to be presented to the operator:

1. Small-scale control systems, where the number of status and counter state data is of the order of 200 or less, and the number of analogue data 100 or less.
2. Medium-scale control systems, where the number of status and counter state data is between 100 and 2000 and the number of analogue data between 50 and 1000.
3. Large-scale control systems, where the number of status and counter state data exceeds 2000 and the number of analogue data exceeds 1000.

Large control systems demand special communication between the operator and the control system, and are often used, for efficient process control on the operator's side, as a set of medium-scale control systems termed 'functional systems'. Small control systems do not usually present any special problem to the operator–system communication, either by the number of data presented or by the number of commands issued.

The basic characteristic functions of medium-scale control systems concerning the reduction of the number of data, the visualization of data and the issuing of commands are:

1. exchange of information between the operator and the process by the supervisory computer or operator station and peripheral equipment;
2. logging of the process data and historic process data performed by the supervisory computer and peripheral equipment;
3. print-out of the process data and the documentation of process states by the supervisory computer and peripheral equipment;
4. the operator's commands issued by peripheral units of the process control system;
5. automatic protection actions performed without the operator's intervention, and the information on these actions presented to the operator.

The main dialogue performance in an expert system is the establishment of firm importance levels and relevance levels of change of crucial process variables, as described in Chapter 14. Thus the confidence in the process result is increased by means of a dialogue process.

8.1 HUMAN DEMANDS

Monitoring and control of production processes involve the following operator's actions:

1. sensory perception,
2. signal discrimination,
3. short-term and long-term memory,
4. mental data processing,
5. decision making on the basis of the data processed and on short-term and long-term memory,
6. some sort of manipulative action to implement certain decisions.

A simple reaction time for the operator is the delay between the occurrence of a single fixed stimulus and the initiation of a response assigned to it. Reaction time T_r to the information presented to the operator is given as (Dallimonti, 1976):

Table 8.1 Reaction times from the information
presented to the operator

Operator reaction times

Brain perception of what the eye sees	0.1 s
Brain recognition	0.4 s
Decision making	4 - 5 s
Action	0.01 - 1 s

$$T_r = a + bH \qquad (8.1)$$

where a is typically 0.25 s, b is typically between 0.25 s^2/bit and
0.33 s^2/bit and

$$H = W \log_2 \bar{A}^2/\bar{E}^2 \qquad (8.2)$$

where H is the input information rate (bits/s^{-1}), W is the input
information signal bandwidth (Hz), $\bar{A}$ is the mean square root of the
change of the input signal amplitude and $\bar{E}$ is the mean square root
of the reading error. Thus, equation (8.1) gives the simple reaction
time of a person to a given amount of input information H. Equation
(8.2) gives the amount of information generated by an instrument or
by a display.

The complex reaction time of the operator is given by the data in
Table 8.1. The estimate of perception, recognition, decision and
action time, T_{PA}(ms), can be obtained from the relation

$$T_{PA} = 270 \ln(n + 1) \qquad (8.3)$$

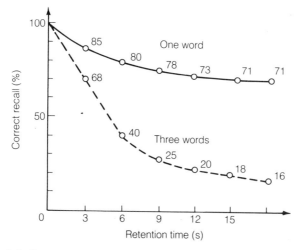

Figure 8.1 Correct recall of one and three words after short time intervals.

where *n* represents the number of logic discriminations required before reacting (Woodson and Conover, 1966).

When a number of instruments and displays are continuously scanned by the operator, the short-term memory of the operator may present serious problems. The short-term memory is the ability of a human to recall information that has just been acquired but has not yet become part of the long-term memory. The work by Peterson and Peterson (1959) indicates how simple words and three words are recalled after short time intervals (Figure 8.1).

This ability of operators emphasizes the need for the simplification and reduction of data presented to the operator. The reduction of process data requires a more efficient coding of useful process information, with the representation of graphic patterns that relate to acceptable and unacceptable process states. The graphic pattern and graphic presentation must reflect the visual and audio perception of the operator. Data on human visual and audio perception, together with relevant data processing functions, are given in Table 8.2.

The ability to work also depends on the physical condition of the operator (see Figure 8.2). The reaction time of an operator is slower under the influence of alcohol or if the operator is alone or on duty during a holiday period.

Table 8.2 Visual and audio perception of the operator

	Visual perception	*Audio perception*
Range of signals accepted	Light 380 nm–780 nm red–violet −10°–+50° from optical centre	Sound 10 Hz–15 kHz
Peak sensitivity	540 nm yellow–green	1 kHz–3 kHz
Resolution	angle minute* 1–1.5 at ± from centre	10^{-16} W/cm²–0 dB at 1000 Hz 10^{-3} W/cm²–140 dB
Information flow through sensors	2×10^6 bit/s	4×10^4 bit/s
Conscious data processing	50 bit/s	–
Reception sensitivity	570 relative different intensities at white	325 relative different intensities at 2000 Hz

* corresponding to a resolution of 0.5 mm at 1 m

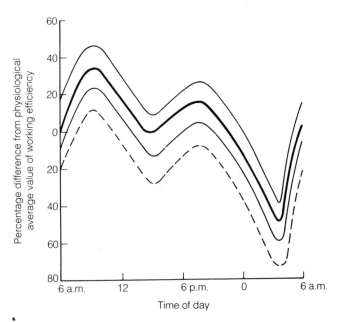

Figure 8.2 Physical condition of the operator.
—— Physiological
—— effective (motivation ± 10%)
—— alcohol (>0.2% alcohol)

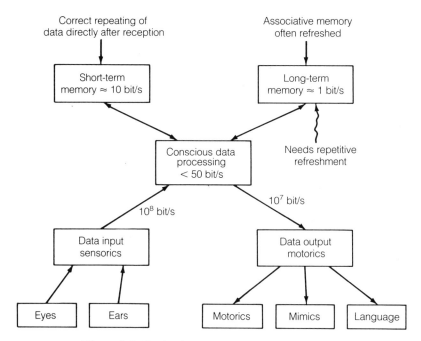

Figure 8.3 The basic pattern of human data processing.

The basic pattern of human data processing can be schematically given as presented in Figure 8.3 where short-term and long-term memory is included. The expert system is expected to assist the human operator with the same time delay as would be expected from human experts. Since the complexity of problems prolongs the decision making process in machines (usually proportionally to the number of rules involved), the solution lies in the fact → rules searching algorithm, which behaves in a way similar to expression (8.3) (Brajak, personal communication).

8.2 THE OPERATOR'S FUNCTIONS AND STANDARDIZATION

In process control applications, the operator–computer communication can be divided into three distinct levels:

Table 8.3 Communication functions and channels for the three control levels

	Process control operations level	*Process engineering level*	*Programming level*
Functions	Process control by exception. Determination of the reason for process deviation. Immediate process action. Acknowledgement for the process action. Data logging. Event recording. Consulting on events from knowledge base.	Gathering of data for process evaluation. Entering constants and control equations. Implementation of application programs and control systems. Creating knowledge base.	Input of system programs. Trouble-shooting. Maintenance. Systems expansion. Development of specific process control algorithm. Development of specific supervisory programs. Program dumping. Bulk reloads of the system.
Communication channels (units)	Display read-outs. CRT consoles. Special purpose keyboards. Work station.	Line printer. Typer. Diskette units. Special purpose keyboards. CRT consoles Work station	I/O type unit. Diskette unit. Printer.

1. process control operations,
2. process engineering,
3. computer programming.

The differences between these applications lie in the speed and frequency of communication, in the method and amount of data presented, and in the level of the user's knowledge, responsibility and education. Typical communication functions and communication channels for these three levels are outlined in Table 8.3, which also refers to their hardware facilities.

The type and relative locations of information devices for the control of processes usually maintain the basic objective of allowing the management of the plant by a single operator or crew in all normal and abnormal situations. In large and medium control systems, two principles are involved to provide this objective:

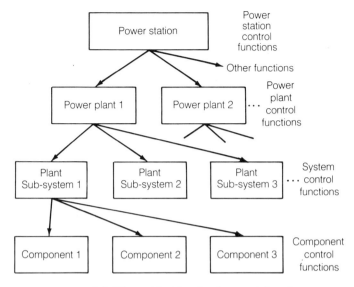

Figure 8.4 Hierarchical levels of control functions.

1. separation of process control functions and all other non-control functions into sub-sets that can be separately observed at any time;
2. hierarchical distribution of discrete control functions into a multi-level organizational structure.

The control levels or priority levels of a power station are illustrated in Figure 8.4. For instance, the control of a power station is handled at the supervisory level, allowing the coordination of power plants (e.g. monitoring the overall station efficiency, maintenance, staffing) and the control of separate power plants includes the control of the entire plant as a system such as the plant start-up and plant load change policy. The control of plant sub-systems allows efficient and quick operations in separate plant sub-systems (e.g. burners, cooling system, heat exchangers) and system component controls allow separate operation of each plant component (e.g. pump, motor, valve).

To assist the operator to undertake the proper action at the correct time, some rules, elements and propositions for the operator–system dialog have been standardized (DIN, 1979), and given

Table 8.4 Some standards for operator–system dialogue (optical signalling)

Main characteristics	Group characteristics	Corresponding alphabets	Recommended max. no. of elements	Comment	Examples of elements
		Alphanumeric character	64 (95) characters	Language dependent	1, A, a,
	Figure	Abstract figures	<20 at same time	Language independent	
		Pictograms	<20 at same time >20 for known pictograms	Language independent	
		Lines	4 different types	–	
Shape		The type of writing	2 different types	–	1, A, a, *1, A, a,*
		Linear distortions	2 different types	–	A, B, C, D **A, B, C, D**
	Form	Figure size	3 different types	Logarithmic rates recommended	1 ^{1} ^{1}
		Line width	3 different types	Logarithmic rates recommended	
		Type of lines	3 different types	–	– – – –, · · · ·, – · – ·,

	Texture	Type of two-dimensional hatching	3 different types	–	–
		Type of colour	Specific colours	6 + black and white	Only for characters from 20′ of visual angle; for smaller angles only red, green, blue and purple	purple, blue, blue, green, red, yellow
		Saturation	–	Not for coding	–
Colour		Light	Light	2 different types	Alternative or redundant coding	green–light green
		Contrast	Colour contrast	2 different types	Colour on coloured background not recommended	Positive (colour on neutral background) negative (neutral on colour background)
			Neutral contrast	2 different types	–	Positive (white or grey on black) negative (black on white or grey)

Table 8.5 Some standards for operator–system dialogue (optical signalling)

Main characteristics	Group characteristics	Corresponding alphabets	Recommended max. no. of elements	Comment	Examples of elements
	Position	Absolute position	9 different positions	–	Upper right Bottom left
		Relative position	2 different positions	–	Indices
Place	Orientation of figures without reference	Orientation	8 different orientations	–	
	Orientation of figures with reference	Position	24 different orientations	More orientations also possible	
	Time variation of shape	Time variation of figure	3 different types	Applicable to process control purposes	Slow, fast, pulsed, stepwise, continuous
Time	Time variation of colour by pulsation	Frequency of colour pulsation	3 different frequencies	Application for danger and attention attraction	0.7 Hz, 2.2 Hz
		Duty cycle	3 different types	–	1:1, 1:3, 1:10
		Speed of change	2 different types	Applicable to process control purposes	Continuous, stepwise
	Time change of place	Translation	–	Specific for process control purposes	Continuous, stepwise
		Rotation	–	Specific for process control purposes	Continuous, stepwise

Table 8.6 Some standards for operator–system dialogue (acoustical signalling)

Main characteristics elements	Group characteristics	Corresponding alphabets	Recommended max. no. of elements	Comment	Examples of elements
Type of sound	Tone	One tone	1	–	Sine tone
		Many tones	–	–	Ground tone with upper tones
	Disturbances	Harmony	2	–	Harmony / Disharmony
		Noise	1	–	Buzzing, ringing
	Speech	Complete sounds	3	–	–
		Call	10	All necessary warning and calling	–
		Words, text	–	Very articulate	–
Frequency	Chosen frequency	Chosen frequency	3	Do not apply octave frequencies	–
Level	Acoustical sound level	Acoustical sound level	3	Minimum 10db above noise	–
	Duration	Duration	2	–	Long, short
Time	Time variation of frequency and tones	Series of tones	2	–	Melody
		Speed of change	2	–	Continuous, stepwise
		Frequency change	3	–	Slow, fast
	Time variation of level	Speed of change	2	–	Continuous, stepwise
		Frequency of change	3	–	Slow, fast
		Pulsation	3	–	Pulses, regular pulsation

in Tables 8.4 and 8.5 for optical signalling, and in Table 8.6 for acoustical signalling in process control systems. The standardization of the operator – machine interface does not yet include the field of intelligent machine behaviour.

9
Recognition and learning

9.1 RECOGNITION AND LEARNING IN OPERATORS

Process operators are trained on the job and schooled, selected and specialized to become familiar with the process and process tools. Traditionally even whole nations have been known as technology promoters, such as the Chinese in porcelain and the Czechs in glass manufacture. Neglecting traditional inclination and schooling, the most important factor is the individual's experience, based on experiments, innovation and trial and error. The essence of this procedure is the selection of several process variables, events or phenomena that are crucial for a given process at a given time/state sequence. Such data are taken from a possibly large number of states and variables. The testing, selecting and internal evaluation is mostly an unconscious process and the result can be explained by the operator as a verbal sequence of procedures and/or actions according to the detected process state. Much help is given to the operator by a refined presentation of process states and trends. As a result of learning there exist:

1. definite probability estimation in operators for certain process states (Bayes estimation);
2. definite rules and hypotheses, value estimations and connections with process events and trends (rules and hypotheses);
3. individual vision of the process in each of its stages of development and implementation (individual process semantics).

The basis of this knowledge lies in the acquired experience and will of operators to refresh and increase their control over the process.
   The operator–system interaction is achieved by:

1. acquisition of initial data for data processing and manipulation;
2. manipulation of data as a real-time dialogue between the operator and the computer system;
3. presentation of data and results.

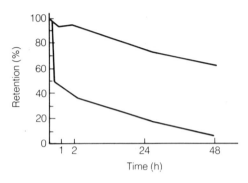

Figure 9.1 Retention of knowledge versus interval following learning.

Most process data are acquired by the computer process interface. These are data on process variables and states, such as temperatures, pressures, flows and voltages. The lowest level process data are transferred to the computer from different existing documents in a computer-readable form, mainly by key-in operations by the operator. These are data on changing conversion units, limit values, timing restrictions and various comments on process behaviour.

The dialogue process consists principally of sending and receiving messages between the operator and the computer via an interface. The quantitative measure of merit of the dialogue can be calculated when the content of information sent by each of them is compared to the content of information received by any of them. There are crucial points in this dialogue design:

1. creation of an environment for the user to use the computer properly;
2. creation of a safe and comfortable dialogue. This implies, for example, the standard keyboard style of terminals, the same special characters all over the plant control system, and dangerous commands to be typed character by character.
3. Involvement of the user in the system and dialogue design process.

The goal of a dialogue is the recognition of process states. The recognition of process states is based on the previous knowledge of the process state interdependencies, as opposed to the recall process by the operator. The result is a higher percentage of retention of the operator's knowledge of process states (Figure 9.1). The success of the dialogue depends, therefore, on a menu-driven dialogue, the

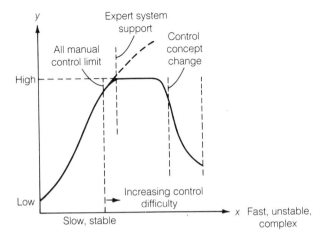

Figure 9.2 Degree of operator control.
y = requirements for operator-system engineering
x = process dynamics

user's involvement in the system and the dialogue design process, and on the available computer graphics.

The design of an operator system dialogue based on Pascal, such as the CCITT language, named MML (man–machine language) (Hornbach, 1982), facilitates the operation and maintenance of computerized process control systems.

A fully successful operator–system interaction can be designed for coarse processes that change slowly with plenty of information feedback. There are environment critical processes that degenerate rapidly and with much interdependent complexity where operator control is impossible, as shown in Figure 9.2. To make operator control possible, an adequate design of basic process units, computer hardware and software must be made (ASEA-ATOM, 1982), including expert system engineering (ABB, 1989).

9.2 THE FUZZY CONCEPT OF LEARNING MACHINES

For control with one input and one output the calculation of control action consists of four steps (King and Mamdani, 1986):

1. calculation of the instant error and change in error;

Table 9.1 Some values of membership functions of fuzzy variables depending on instant process error (PO = positive zero, ...)

Change in process error: 2 , 1 , 0 , −1 , −2

Instant process error

	−6	−5	−4	−3	−2	−1	0	1	2	3	4	5	6
PB	0	0	0	0	0	0	0	0	0	0.1	0.4	0.8	1
PM	0	0	0	0	0	0	0	0	0.2	0.7	1	0.7	0.2
PS	0	0	0	0	0	0	0.3	0.8	1	0.5	0.1	0	0
PO	0	0	0	0	0	0	1	0.6	0.1	0	0	0	0
NO	0	0	0	0	0.1	0.6	1	0	0	0	0	0	0
NS	0	0	0.1	0.5	1	0.8	0	0	0	0	0	0	0
NM	0.2	0.7	1	0.7	0.2	0	0	0	0	0	0	0	0
NB	1	0.8	0.4	0.1	0	0	0	0	0	0	0	0	0

Fuzzy variables

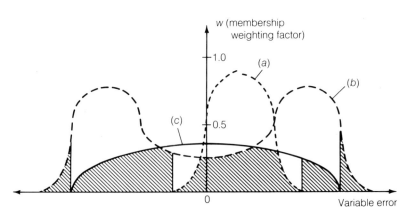

Figure 9.3 Max/min principle applied to three membership functions of the same variable error.

2. conversion of error into fuzzy variables;
3. development of defined decision rules with the help of compositional inferring rules such as:

$$\text{if } X \text{ is } A_i \text{ then (if } Y \text{ is } B_i \text{ then } C_i) \tag{9.1}$$

or as expressed with the membership function:

$$\mu_c(x, y, z) = \text{Max Min}(\mu_{A,(x)}, \mu_{B,(y)}, \mu_{R,(x,y,z)}) \tag{9.2}$$

where C is the control action and the membership function of the proposition R is calculated from input, output and control membership functions as:

$$\mu_R, (x, y, z) = \text{Min}(\mu_{A,(x)}, \text{Min}(\mu_{B,(y)}, \mu_{C,(z)})) \tag{9.3}$$

4. calculation of the determistic set-point for process control.

Table 9.1 gives the definition field and values of the membership functions of fuzzy sets for a concrete indicated process error.

The extension of the definition field, while introducing other error values, is also indicated. When observed in a continuous spectrum of the final action as depending on the error state, the membership function realized from equation (9.3) applied to three fuzzy sets is given in Figure 9.3.

Generally, the machine learning procedure when based on fuzzy set theory consists of the change in the membership function weighting factor. When, for example, the negative big (NB) function value of 0.8 for the instant process error of −5 from Table 9.1 does not produce an adequate effect, then a value of 0.9 or probably 1 should be appointed to this element. The problem lies in what changes should be made to other elements of the NB function and even of the NM (negative middle) change when the unsatisfactory behaviour of the element has been observed. Some answers to these questions are given in Chapter 13.

REFERENCES FOR PART THREE

ABB (1989) *Graphical dialogue environment.* Publ. No. D CRH 1333 89 E.

ASEA–ATOM (1982) Special Features of the Control Equipment for the ASEA–ATOM BWR. Appendix to *IAEA Guide Book on I&C.*

Brajak, P. (1990) Personal communication.

Brillouin, L. (1956) *Science and Information Theory*, Academic Press, New York.

Dallimonti, R. (1976) *Instrumentation Technology*. May, 39–44.

DIN 66234 (1979) CRT display working place, coding of information, April.

Hornbach, B. (1982) *IEEE Trans. Comm..* **30**, No. 6, 1329–36.

King, P.J. and Mamdani, E.H. (1986) The application of fuzzy control systems to industrial processes, *Automatica*, **13**, 235–42.

Kononenko, I., Bratko, I., Roskar, E. (1986) ASSISTANT: A system for inductive learning, *Informatica Journal*, **10**, 1.

Peterson, L. and Peterson, M. (1959) *J. Exp. Psychol.*, 193–8.

Puendtler, K. (1977) *Elektrizitätwirtschaft.* **17**, 579–85.

Winograd, T. (1987) Artificial Intelligence **31**, 250–61.

Woodson, W.E. and Conover, D.W. (1966) *Human engineering guide for Equipment Designers*. University of California Press, Los Angeles.

Part Four

Action

While the first two parts cover the major aspects of process control from the designer's point of view, the third part gives the foundation of the dialogue process. This fourth part treats the semantic aspects of process action, describing two crucial notions for process control in intelligent devices: the importance and relevance of information. Prior to this analysis , an overview of process action tools and procedures is given.

Automatic knowledge acquisition in real-time expert process control is presented as an illustration of the method. The algorithm has been applied to the power plant data. The foundation of the presented method is based on the modified matrix model of automatic process knowledge acquisition and a different fuzzy set control description that always allows a tree-like decision procedure using only one selected branch and optimum solution. The actual process state message is obtained from the mesage database, being dependent on the semantic evaluation of process variables.

10

Tools for process control

10.1 CHANGING PROCESS STATES

Once the problem of hardware for process control had been solved, the next move, in the 1970s, was to provide monitoring and simple control systems. Inevitably, things have advanced since then, and one definite trend is towards integrated cooperative automation of specific processes. Such automation is absolutely necessary for fast and complex processes (e.g. electric power generation, transmission and distribution systems or natural gas distribution systems) and consists principally of two possible system behaviour steps:

1. recognition of the situation for automatic small-size system adaptations and the small-size control action and automatic reporting on the action to the operator;
2. automatic on/off control action on the recognition of a specific process state, such as the alarm state or danger and reporting to the operator on the actual on/off control actions performed.

The second step is performed on the basis of process control algorithms simpler than the first action, which requires fast, accurate and updated action, model calculation and model verification algorithms because it usually represents fast and complex action on the process behaviour. A relatively simpler case of the first step can be executed for slow processes where the recognition of the process action is modelled and reported to the operator who then decides and/or undertakes the appropriate action.

Four cases of process control actions considering the role of the operator are given schematically in Figure 10.1.

In Figure 10.1 case (*d*) sometimes allows automatic actions over larger ranges of variables than the previous cases (Fogarty, 1989). This can be done using different sets of system rules. Some rules are designated for optimization purposes, some for the control range and

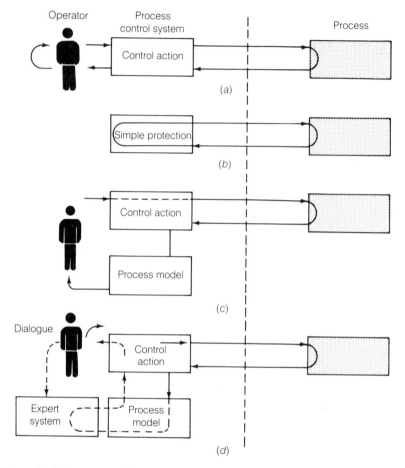

Figure 10.1 Four cases of process control action: (*a*) control action by operator's decision; (*b*) automatic, by control equipment; (*c*) control action by operator's decision based on process modelling; (*d*) automatic, by control equipment with feedback of information to the operator through the dialogue between operator and expert system.

some for fault diagnostic and safety reasons. The goal of the actions performed is to search for the optimization region within a minimum time using different rules from the control range, or reaching the control or optimization range from the false diagnostics and safety region (Figure 10.2). The borders of the regions and rules are defined by experienced operators and plant engineers. Rule modifications, the performance evalation of modified rules and verifica-

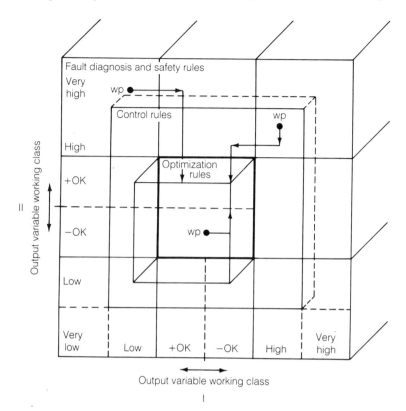

Figure 10.2 Action classes in process control – the case of two output variables (wp: working point).

tions can be done automatically and they are treated in Chapters 12 and 13.

10.2 OPERATOR/PROCESS INTERACTION – ACTIONS, COMMANDS AND TIMING

The interaction of an operator with the process can be caused by three main reasons as viewed from time requirements (Table 10.1):

1. Fast interaction (e.g. start/stop or on/off) operations on final control elements, called control operation; the time for these operations ranges from a few seconds to a few hours.
2. Slow interaction, called the planning operation (e.g. load and

Table 10.1 Data and functions for different types of operation

Type of operation	Content of operation	Data and functions needed for operation
Control operation ranging from a few seconds to a few hours	On/off operations on final control elements	Process status signals Process command signals
	Start/stop operation on BPU	Process status analog and count signals BPU local console
	Auto/manual operation on BPU	Process data acquisition and presentation Limit values monitoring auto/manual facilities
	Increase/decrease operation on regulators	Analogue and status data monitoring Limit values monitoring set-point issuance
Planning operation ranging from a few minutes to a few days	Operation scheduling Load and demand prediction Economic planning Operation simulation Local process optimization	Process data acquisition Monitoring of limit values Process modelling Process security modelling Load and demand modelling and stratification Process economy modelling
Following up ranging from a day to a few months	Daily, weekly and monthly production logs Load and demand reports Event reports Process environment reports	Production, load, demand and event data Process environment data

BPU – basic process unit

demand prediction, process simulation for optimization purposes, economic production, distribution and process security); the time range for these operations extends from minutes to a few days.
3. Very slow interactions, called following-up (e.g. operations caused by daily, weekly or monthly change in production, demand, load, power and material exchange); these operations range from one day to a few months.

The main tasks of various computer control systems at different hierarchical levels of control are given in Table 10.2. The control levels below a sub-station or remote station are usually equipped with synoptic panels with a dedicated operator–system communica-

Table 10.2 The main tasks of different computer control systems

Level of hierarchy	Operation	Planning	Following up
National grid	Production	Consumer prediction	Reporting and accounting statistics
Interconnected utilities	Supervision of: consumers production reserves process network Operation and control of plants	Production schedules Balance planning Planning of reserves Coordination of overhauls	Following up of efficiency Fault analysis
Utility	Production	Consumer prediction	Reporting unit accounting statistics
Regional grid	Supervision of: consumers production reserves process network Operation and control of utility	Production schedules Balance planning Planning of reserves Coordination of overhauls	Following up of efficiency Fault analysis
Group of plants	Production	Short-term planning according to directives	Reports on production Accounting data Statistics
District grid	Supervision of: consumers production process state plant components Operation and control of sub-station		
Plant	Control of variables, sequential start/stop functions, automatic system restoration, protective functions, supervision of process variables, auto/manual for local equipment	Work planning	Sequential event recording

tion and used both for synoptics of the basic process unit and the process controller.

The whole operator–system dialogue and operator interaction with the process is supported and designed on the following basis:

1. system and application software, including the database/knowledge base system;
2. hardware data acquisition, processing, presentation and hardware command execution;
3. operator–system dialogue elements, as mutually agreed upon.

The basic components of a programming system supporting the operator–system interaction are system functions executed in real-time and extended real-time, and the user's application functions which are all usually supported by the following program functions in real-time:

1. data acquisition, data processing and visualization of process data;
2. data acquisition, data processing and visualization of control system data;
3. process control including the modelling and real-time expert system;
4. control of the control system.

Software support for extended real-time functions enables process planning (i.e. the change of system parameters such as the change of process, process control and visualization databases), reporting (i.e. the acquisition of process data and their systematic presentation in given time intervals or at operator's demand) and extended real-time processing (i.e. all data processing functions longer than the given system refreshing interval).

The application programs specially developed by the system user enable the fulfilment of the user's specific functions, appropriate, for example, for their process hardware and technology. They are limited by the integrity and speed requirements imposed on the real-time and extended real-time software models.

Application programs use and share the same computer resources and are therefore limited by the integrity and speed requirements imposed on real-time and extended real-time software models.

The appearance of a process event indicates the change in a database and is processed according to the content attributed to its appearance.

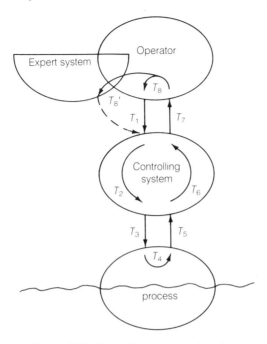

Figure 10.3 Overall system reaction time.

Events demand different simple data processing, for example, a list of events, alarm and failure recording, as well as chronological event recording.

A composite data processing of events results in display refreshment, which supplies information to the operator on events such as display of real-time mimic elements, grouping of real-time mimic elements, and selection and authorization of work of operators.

A list of events, alarms, failures and chronological events are issued on printers for the operator's information.

The operator–computer communication can impose special requirements in real-time system operations. Here, the overall system reaction time is essential for the system functioning (Figure 10.3). Table 10.3 outlines individual reaction times for trained operators, as depicted in Figure 10.3.

The interaction of an operator with the process is done by changing the process control functions; but there are many factors that influence these functions (Wilhelm, 1979); they are outlined in Table

Table 10.3 Data for Figure 10.3 for trained operators

Reaction time	Approximate time range	Brief description
T_1	0.4 s–10 s	Execution of required control or dialogue action of the operator on control system (on keyboard, functional keyboard or panel command push-buttons)
T_2	0.1 ms–30 s	Execution of accepted command toward a final control element; the main time delay can be caused by communication facilities
T_3	0.1 ms–100 s	Execution of command at the final control element; the main time delay can be deliberately set high enough to enable system stability
T_4	up to 1000 s	Process reaction; ranges from a few minutes to tenths of minutes depending on process type
T_5	0.1 ms–10 s	Process state detection time; depends on sensor reaction time
T_6	0.1 ms–10 s	Control system event processing; depends largely on I/O filter constants and data processing algorithms
T_7	10 ms–10 s	Visualization of process event to operator
T_8	(individually) 0.5 s–30 s	Operator reaction time to process event and decision to take immediate action in real-time control environment
T_9	(individually) 0.8 s–20 s	Operator reaction time to process event and decision to take immediate action in real-time environment supported by an expert system software

10.4. The usual main process control commands for the operator–system communication are:

1. start command, which requires the transition of the process from the OFF to the ON state;
2. stop command, which requires the transition of the process from the ON to the OFF state;
3. activate command, which requires the transition of the process from the inactive to the active state;
4. deactivate command, which requires the transition of the process from the active to the inactive state;
5. unlock command, which requires the transition of the process from the unavailable to the available state;

Table 10.4 Factors that influence process control functions

Factor	Factor's influence	Comment
Operator Final control element	Operator inputs status of final control element	In auto/manual systems In dynamic or erroneous situations
Sensor	Status of sensor	In dynamic or erroneous situations
Control functions	Status of related control functions	Depending on mutual connections of basic process units
Stages of operation	Status of production in batch or sequential processes	Changes in control structure and functions
Process state	Change of process control function	Some regions of operation may require changes in control functions

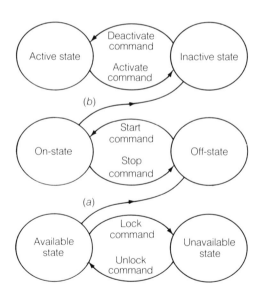

Figure 10.4 The basic process states transitions and process commands; (a), (b):
optional transitions.

Table 10.5 Specific features of process control equipment for command and process signalling

Input or output device action	Communication mode	Operator–system input/output expected response times
Operator → QWERTY keyboard	Text input mnemonics	0.2 s/character
Operator → QWERTY keyboard	Coded input	0.2 s/character
Operator → Functional keyboard	Functional keys	0.2 s/key
Mimic display → operator	Question	1 s/question
Mimic display → operator	Warning on false response	0.5 s/warning
Computer action	Computer action	0.5–10 s
Mimic display → operator	Information on action result	1 s–10 s
Mimic display → operator	Information on spontaneous events	0.1 s–5 s

6. lock command, which requires the transition of the process from the available to the unavailable state;
7. notice command, which requires the process control system to recognize a state transition and to take any necessary action.

The basic process state transitions and process commands are given in Figure 10.4.

Particular process states (e.g. an unavailable state) can be reached by issuing commands and by changes in the process equipment and software behaviour due to previously unpredicted errors and failures. In such cases, after the system has been repaired, particular process states are again attained by operators or service staff issuing auxiliary operations and manual commands. The procedure whereby an operator intervenes after such a failure to put the system into a particular process state should be well defined and carefully studied; it demands the manual preparation of all basic process units, so that the changeover can be successfully made to the computer control operations. This presents special difficulties for some plants, especially where conditions are hazardous for humans or the units are not easily serviceable. The corresponding commands and process signalling can be issued on different process control equipment. The specific features of different process control equipment for command

and process signalling are given in Table 10.5. The indicated time intervals are heuristic and correspond to expected reaction times of the operators and computers.

The control of process variables is performed either as an on/off control, or as a regulation or governing function. The on/off control is a procedure that begins with the selection of the mimic element. The computer checks whether the operator controls the selected process element by obeying the following suppositions:

1. control competence must be attached to the operator;
2. process element data must be refreshed;
3. on/off control function must be allowed;
4. only this control point is in the control procedure for the selected process element;
5. the logic control function must be fulfilled.

The operator is informed about the result of the computer check procedure and, if the computer permits the on/off control, the appropriate keyboard action may be executed. The keyboard is blocked until the time supposed for the execution of this process command elapses, or until the acknowledgement of the command execution. The change of the process element state is usually registered and, if it does not occur in the prescribed time interval, an alarm is issued to the operator.

A regulation or governing function is a procedure that must fulfil the same suppositions as those for the on/off control. The computer issues permission to the operator for the execution of the regulation, when two procedures become possible:

1. increment regulation, which is executed by the function keys that allow positive or negative increments of the process final control devices;
2. set-point regulation, which is executed by the operator typing in the set-point value.

The new process element value is checked against the given limit values. Both instantaneous and previous values are displayed.

10.3 SOFTWARE TOOLS AND ACTION

Software tools for the expert process control include different and usually distributed features such as:

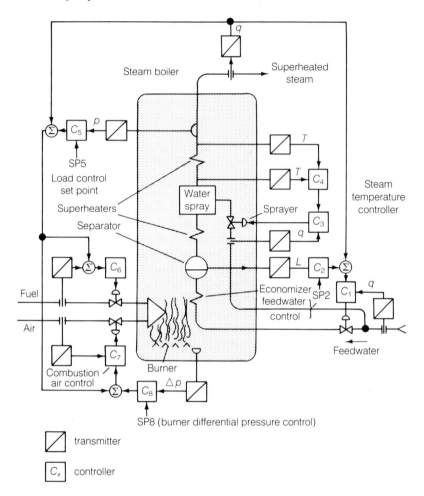

Figure 10.5 Process instrumentation scheme of a simple steam boiler. q: flow; $\Delta p, p$: pressure; L: level; SPx: set point value; Σ: summation point.

1. local intelligence in smart transmitters and final control elements which allow fast recalibration, diagnosis and change of process parameters;
2. network support for interconnection of different process control computers and interaction to managerial level computers, such as offered by UNIX releases (Brajak, 1990)
3. expert system tools as presented in Chapter 1, where the effective rule generation, rule based control, manual operation, rule

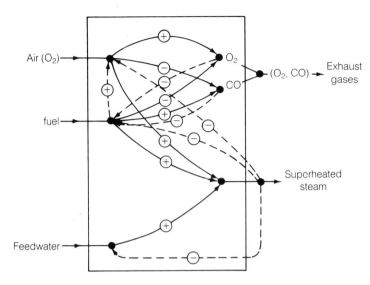

Figure 10.6 Main process and control interactions in a steam production unit – a qualitative scheme.

—— = process interaction
------ = control interaction
$\oplus, \ominus$ = increasing or decreasing action

learning, process interface, process simulation mechanism and simulation builder are integrated (Fogarty, 1989).

Example: action in an expert system environment
The scheme of a simple steam boiler and attached control circuits is given in Figure 10.5. There are two main control circuits:

1. the steam control, where the production of the superheated steam is connected with the fuel/air consumption through controllers C6 and C7;
2. feedwater level control in drum controllers C2 and C1.

The control of steam temperature is performed through controllers C3 and C4 and the steam pressure control is obtained through the C5 controller connected to the fuel/air control. When all control circuits are tuned to the process and there are no changes either in the input variables or in the process itself (such as calcination deposits in pipes, burner malfunctions, and valve inoperability), the presented

Table 10.6 Functional spectrum of SIMEXPERT S5 expert tool for SIMATIC S5 process controller from Siemens

Controller type	Functions				
	Diagnosis		*Parametering*		*Putting into operation (checklist)*
	On-line	*Off-line*	*Hardware*	*Software*	
SIMATIC S5 115 U	X	X	–	–	X
SIMATIC S5 135 U	X	X	–	–	X
SIMAIC S5 150 U	X	X	–	–	X
IP 241	–	–	X	X	–
CP 524	–	X	X	X	–
CP 535	–	X	X	X	–

– does not apply
X applies

control scheme can cope with the task of control producing the pre-set value of steam. In the event of processes or control changes which are not 'pre-wired' in the control structure, the operator has to take control, switching over a particular controller to manual control to govern the process in its optimum way. Thus the effects of, say, negligence by the third shift operator or inability to govern the system can lead to unexpected losses of and damage to the hardware and potential danger for personnel and the environment.

A more complex interaction structure of input and output variables is given in Figure 10.6. Here the influence of exhaust gases is also added to the previous variables from Figure 10.5. The presentation in Figure 10.6 is given in a qualitative manner, thus omitting unnecessary complications. It is worthy to notice the balance of material and energy in both schemes of the presented steam production unit.

There are several actions that can be attributed to the potential expert system for the control of steam production:

1. information to the operator on the system trend, both long-term and short-term with an indication of the most interesting trends;
2. information to the operator on possible failure points and prevention measures to avoid them;

3. small-scale corrective action in control loops regarding the controller's constant adaptation to the process state and information to the operator on parameter changes;
4. recommendations to the operator on actions in case of $n - 1$ failures in the process equipment.

Actual expert system tools for the Siemens programmable logic controller family are given in Table 10.6. As can be seen, the tasks of such an expert system are intended mainly as a control structuring tool. On-line diagnosis gives data on the cause of the STOP state in either hardware or software and leads to a solution in several cycles of the dialogue. Parametering includes graphical features for the pre-setting of components (DIL-switches, short circuits, wiring tables, programming data on S5 components). The overall frame functions include control of intermittent inputs, a help function, protocol issue, documentation and 'hot-line' connection to the service.

11

Process protection

11.1 INTRODUCTION

The working body (system) is the process body observed separately
from all the other mutually interrelated bodies (systems) whereby
they are all treated as the system environment. The process is an
outer macroscopic reflex of variable and microscopic changes in the
working body. The process states can be distinguished as stable and
unstable states, according to their character. By isolating the working body from the system environment, stable states are obtained for
the majority of processes.

The interaction of the environment inevitably involves the instability of the process. Thus there are two opposing tendencies during
the process life; the outer tendency from the environment usually
works towards process instability, and the inner tendency inside the
system usually works towards process stability. Since almost all
systems are bound by many sub-systems with their environment,
most systems are inevitably unstable.

11.2 THE PROTECTION MECHANISM

To prevent the effects of process state instability, the protection of
particular processes is organized according to the best known experiences and calculations for each separate process. It is useful to
distinguish between active and passive protection, and between static
and dynamic protection. Figure 11.1 shows the interrelations of three
systems. Consider, for example, system 1 in Figure 11.1 as a boiling
water system, system 2 as a saturated steam system and system 3 as
part of the water circulating tubes and armatures. Then the following
apply:

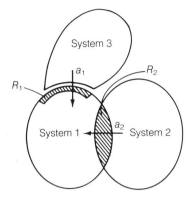

Figure 11.1 Active and passive protection measures:
a_1 = action of System 3 on System 1
a_2 = action of System 2 on System 1
R_1 = passive protection measure of System 1 against action a_1
R_2 = active protection measure of System 1 against action a_2

1. passive protection of system 1 against the action of system 3 on system 1, for example a better isolation and screening of tubes to obtain improved mechanical performances when conducting boiling water;
2. active protection of system 1 against the action of system 2, for example the inclusion of the facility for forced cooling of the saturated steam.

The active protection facility can either be of a static or a dynamic nature depending on the type of process or process behaviour it covers. A static protection system is used for the protection of process steady work and operates on the basis of stationary process protection variables. A dynamic protection system is used for the protection of process dynamics and operates on the basis of dynamic process protection variables.

Usually only some simple processes start up with manual intervention by the operator. More complex processes have to be automatically started, restarted, put into action and stopped. Both manual and automatic operations can be, or sometimes have to be, supported by dynamic protection systems. The reason for this lies in the complexity and number of decisions that can otherwise be expected from the operator, thus saving time, energy and materials, as well as avoiding dangerous situations for people and process

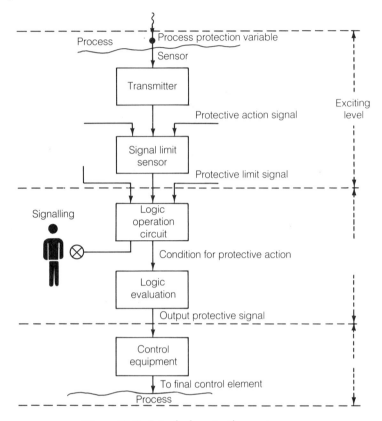

Figure 11.2 A typical protection system.

equipment alike. Automatic actions for process protection in the dynamic range are supported by those actions that serve for the process protection in the steady state. These two types of protection are usually separated in their actions because, when used, the variables and process conditions differ considerably. The typical protection system of either a static or dynamic type consists principally of the same type of hardware and software as given in Figure 11.2 for a common protection channel. The process protection variable is detected by a sensor and then transmitted as a protective action signal to a limit signal sensor. The cause of an appearance of a process protection variable in the range that is of interest for the protection action can be:

1. stochastic change in the process variable;
2. systematic change in the process variable;

3. change in the process variable caused by a change in another part of the process;
4. change in the process variable during repair, damage or unit shutdown.

The protective limit signal generated at the output of the signal limit sensor is fed to the logic operation circuit where it is combined with other protective limit signals to generate (i) appropriate signalling to the operator and (ii) signalling of the protective action condition. The signalling of the protective action condition is evaluated by a redundant logic circuits (e.g. 'two out of three' circuits, or similar) and an output protective signal is generated to the control system which generates the appropriate output protection signal with the required power and time requirements to the final control element.

11.3 AUTOMATIC ACTIONS AND CONNECTION TO THE EXPERT SYSTEM

Protection systems are designed for various processes. The criteria for their design include the following minimum requirements:

1. the existence of process conditions that require protection;
2. the existence of a process variable to be monitored to provide protective action;
3. exact limit values and levels for each process variable that requires protection action;
4. the margin between each protective limit and the level considered to designate an unsafe operation;
5. protection performance requirements, such as the protection system response time, the system accuracy and ranges of process variable conditions (normal, abnormal, accidental) both in magnitude or/and rate (Figure 11.3).

Protection systems are usually implemented by the separate hardware that ensures process protection as specified in the process protection requirements and they are an integral part of process technology. For larger process units, such protection mechanisms can be implemented in expert systems specially when considering the process safety aspects (Ahrens, 1987; Wittig, 1987). The process safety aspect takes into account the totality of the process and its environments, i.e.

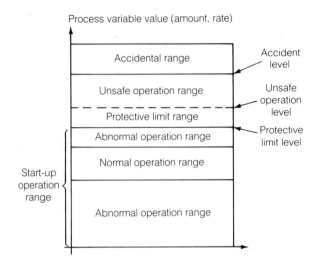

Figure 11.3 The ranges and levels of process variables.

1. failure effect analysis (DIN 25448);
2. risk analysis;
3. operator failure analysis;
4. operability analysis (PAAG method).

These analyses encounter qualitative parameters in the decision procedure, thus forcing the use of the expert system tool. Operator-induced failures are, for example, divided into the following classes of action:

1. too early/too late;
2. careless;
3. not executed;
4. erroneous;
5. in the wrong direction;
6. in the wrong place.

Knowledge bases can be used directly in graphic methods such as the analysis of the disturbance flow (DIN 25419) and the error tree analysis (DIN 25424).

12
Semantics and reaction

12.1 THE SEMANTIC VALUE OF INFORMATION

The semantic measure of information determines the quality of information. Thus, the content of the process mimic can have a negligible quality of information compared, for example, to one row of the event recorder print-out under given circumstances.

The binary digit or bit is the measuring unit for the information quantity and is mathematically exactly defined in the statistical sense. The qualitative estimation of information contains more or less subjective dependent parameters.

Figure 12.1 presents the semantic value of information as viewed through its usability. The ordinate presents the level of the semantic value of information and the abscissa its usability. The four quadrants represent:

quadrant I: important information, usable and precious;
quadrant II: precious information but not usable, not 'at hand';
quadrants III and IV: meaningless, unimportant, empty, false and
   harmful information.

Let us define the unit for measuring the quality of information as the SIT (semantic information bit). For the purpose of the quantitative determination of information, five parameters are defined (Gitt, 1986):

1. semantic quality, q
2. importance, r
3. actuality, a
4. reaction to information, z
5. existence, e

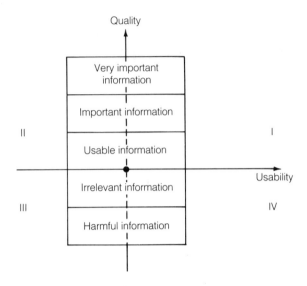

Figure 12.1 Quality and usability of information.

The named parameters involve a certain subjectivity factor which must be taken into consideration mostly by the information user.

Semantic quality, q, measures the meaning or content of information. Its interval is $(-1, 1)$, meaning that 0 occurs when the information is empty or senseless and 1 means that the weight of a positive meaning is maximum.

Importance, r, measures the purpose of information and is closely connected to the attainment of a given goal. It is a very relative parameter; $r = 0$ means that the information is unimportant and $r = 1$ means that it is maximally important.

Actuality, a, measures how recently the information is obtained; $a = 0$ means that the information is not actual and $a = 1$ means that the information is just newly arrived.

Reaction to the information, z, measures the ability of the information user to respond to the obtained information, ranging from being unable to understand the message to the situation in which the user's reaction is disabled, and to the situation in which the user can fully react to the obtained information.

Existence, e, depends only on the information exchange between sender and receiver, being from $e = 0$ when there is no response on the transmitter side to $e = 1$ when the response is fully presented.

Representing the five parameters as a pentagon of area A, Figure

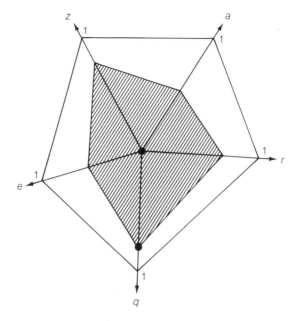

Figure 12.2 Area representation of the semantic quality value. e, q, r, a and z are semantic quality parameters.

12.2 gives a measure of the semantic content. The corresponding relation for the area of pentagon A is

$$A = 1/2 \sin 72° \ (ra + az + ze + e|q| + |q|r) \qquad (12.1)$$

The maximum area is equal to $A_{max} = 2.377641$. The number of SIT should be proportional to the area. In order to circumvent the case when any of the parameters is equal to zero and the area is not, the auxiliary function F is defined as

$$F = \begin{cases} 0 & \text{if} \quad q, a, r, z, \text{ or } e = 0 \\ 1 & \text{otherwise} \end{cases} \qquad (12.2)$$

The sign of parameter q can formally interfere with the area value. In order to give a sign to the whole expression (and meaning as well) the relation (12.1) is transformed to

$$\text{Semantic value} = 20 \ (ra + az + ze + e|q| + |q|r) \ F \text{ sign } (q) \ (\text{SIT}) \qquad (12.3)$$

The semantic value of information can thus take a value between 100 and -100.

12.2   SEMANTIC EVALUATION OF REAL-TIME PROCESS DATA

Process data are acquired in a real-time procedure. There is a problem of the semantic evaluation of these data in process control, at least in process surveillance. Therefore, the given five semantic parameters should be modified to the world of real-time data acquisition of the process variables.

12.2.1   EXISTENCE PARAMETER

Process data are acquired from the process transmitter. The connection can be tested and evaluated for proper functioning, and hence the value of the parameter e can be either $e = 0$ when the connection is out of order or $e = 1$ when the transmitter is on-line. Estimation of the process signal existence can be obtained even from hardware circuits, although it can be performed in software, for example, when putting a process part out of operation.

12.2.2   SEMANTIC QUALITY PARAMETER

Data acquisition for process control is usually organized so that many process variables are sampled. Due to disturbances or failures a piece of information from the process variable can be obtained that does not correspond to other variables in the process. Such information can be taken as meaningless or harmful and should not be taken into account. Control of this parameter can be performed very simply as, for instance, comparing the time variance of the actual sample with the previous one.

Thus $q = 0$ means that the variable value is inappropriate and $q = 1$ means that it is appropriate for further processing.

The registration relevance of the process variable can help in failure and error detection, since any process failure will result in the instantaneous or steady change of one or more process variables. The follow-up of these changes can help in the detection of unallowable process states or failures.

An example of the relevance calculation of variable $V_i(t)$ from variable difference $|\Delta V_i|$ in successive sampling intervals $t, t + \Delta t$, for, t the case of its dependence on variable amount V_i and the parameter t is given in Figure 12.3 (Zupanec, 1990).

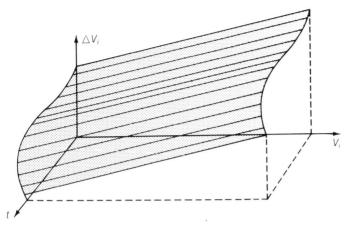

Figure 12.3 The permitted variable difference ΔV_i depending on variable value V_i and the time parameter – the decreasing shape of the ΔV_i with the time parameter is due to the heating effect in the actual process.

12.2.3 ACTUALITY PARAMETER

Actuality is an attribute of all process variables in process control, i.e. they are evaluated through the parameter or importance, since there exists no actual but unimportant process information. Therefore, the actuality of the process variable will be always estimated by the importance parameter.

12.2.4 IMPORTANCE PARAMETER

Any number of the process information that exists and is relevant has to be evaluated from the aspect of importance. The importance is evaluated through the relation of the process variable and its limit values, including the variable amounts and time margins. Thus, the limit value can be a constant, a limiting curve depending on another process value, a limiting surface depending on two other process values, or a limiting space if depending on two or more other process variables and/or the time parameter.

The process value can be unimportant or important, but some necessarily have higher importance levels, depending on the importance of its change in the limit ranges. The importance levels and

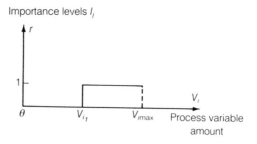

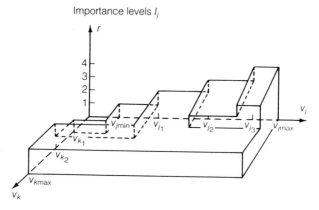

Figure 12.4 Importance levels of process variables. v_i, v_j and v_k are process variables.

limits for three process variables are presented in Figure 12.4. The importance levels, limit values and relations for a particular case are all set up by process experts.

Another possible way to attach importance levels to the process variables from their probability distribution function will be treated in Chapter 13.

12.3   THE CALCULATION OF SIT FOR PROCESS VARIABLES

The quality of the process information of variable V_i is given with the set $Q_i = \{e_i, q_i, r_i\}$, where the parameter r includes the previously defined relevance parameter, while q and e are binary values. Thus, for any of them being zero the quality of the information V_i equals zero. In contrast, when $e = q = 1$ then the value of r determines the information quality, i.e. SIT $(V_i) = r_i$.

N independent process variables with their r values can be taken as the ordered set of number B i.e. (Zupanec, 1990).

$$B = r_1 r_2 \ldots r_N \tag{12.4}$$

where the ith number has the number basis R_i i.e. the resolution of the variable V_i ($R_i = r_{imax} + 1$). When converted to the decimal number, one obtains the decimal equivalent of the process SIT value as

$$\text{SIT} = r_1 + \sum_{i=1}^{N} r_i \left(\prod_{j=1}^{i-1} R_j \right) \tag{12.5}$$

The maximum number of SIT is given by the relation

$$\text{SIT}_\text{M} = \prod_{i=1}^{N} R_i - 1 \tag{12.6}$$

Example
For the three independent variables from Figure 12.4 and supposing that $e_i = e_j = e_k = q_i = q_j = q_k = \text{TRUE}$, $r_i = 1$, $r_j = 0$ and $r_k = 4$ then ($R_i = 2$, $R_j = 3$, $R_k = 5$):

$$\text{SIT} = 1 + (3 \times 0) + (4 \times 3 \times 4) = 49$$

$$\text{SIT}_\text{M} = (3 \times 4 \times 5) - 1 = 59 \tag{12.7}$$

12.4 SEMANTICS AND ENTROPY OF PROCESS INFORMATION

The procedure of calculating the entropy of process variables is presented in Figure 12.5. The corresponding entropy levels are as follows (b_i = number of bits of the variable V_i):

(i) total entropy of all N independent process variables

$$H_1 = \sum_{i=1}^{N} b_i \tag{12.8}$$

(ii) entropy after elimination of J non-existent variables

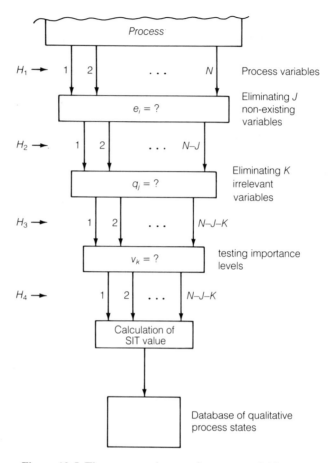

Figure 12.5 The entropy change of process variables.

$$H_2 = \sum_{i=1}^{N-J} b_i \tag{12.9}$$

(iii) entropy after eliminating K irrelevant variables

$$H_3 = \sum_{i=1}^{N-J-K} b_i \tag{12.10}$$

(iv) entropy after testing importance levels

$$H_4 = \sum_{i=1}^{N-J-K} \log_2 (R_i) \tag{12.11}$$

Thus the entropy of the process is decreased after each level of data processing. By a linear decrease of the number of importance levels a further exponential decrease of entropy is to be expected according to equation (12.11).

12.5 SEMANTICS AND REACTION IN PROCESS CONTROL

According to recent research by Terry Winograd, the meaning of the communicated message is in the interpretation assigned to it by the listener. Thus, any semantic of process messages should be left to the operator. Still there are process situations demanding action in the machine which include some type of machine interpretation. This interpretation is necessarily limited to the notion of the information quality. Whether introduced by the operator or by a machine the action tends to put the process into another production state which contains another semantic value of its variables. If it is driven toward more secure values of the variables, then the semantic value of the process decreases because the importance parameter of the variables decreases. If the operator directs the process toward less safe process states (toward variable limits), then the process semantics increases.

The number of semantic units (SIT) calculated according to equation (12.4) represents the coded qualitative state of the process. In order to be usable to the operator, this number has to be decoded into meaningful content. One possible way to present the qualitative state of the process in real-time by means of the SIT value is to use a linguistic base of qualitative description of the process state (reaction) that can contain either information to the operator on the meaning of the actual process state or information to the process operator or manager on what action to take in a certain process state.

The reaction base is therefore a set of ordered information $I_1 \ldots I_m$, M being the maximum number of SIT, in which all necessary descriptions of the qualitative process state are collected. This base will be as good as the quality of the relevance parameters and importance levels of all process variables – provided they are independent. Thus, a multi-dimensional space of process variables is reduced to a one-dimensional database of process reactions. By an appropriate reduction of the importance level number, such a database can be of rather reduced size and complexity.

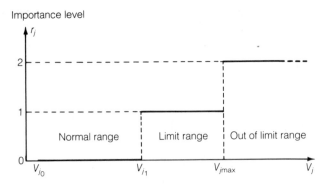

Figure 12.6 Importance levels of the armature voltage variable V_j of the DC induction motor.

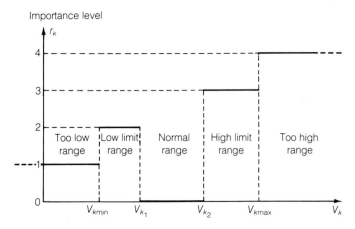

Figure 12.7 Importance levels of the angular velocity variable V_k of the DC induction motor.

Example
When taking variables V_j and V_k from Figures 12.6 and 12.7 respectively as real variables of the DC induction motor where V_j is armature voltage, and V_k the angular velocity ω, then, according to equation (12.6) the maximum number of SIT will be $M = 3 \times 5 - 1 = 14$ and the linguistic base will contain 15 messages:

I_1: empty
I_2: 'High armature voltage; do not increase it!'

I_3: 'Too high an armature voltage; immediately decrease it!'

.

.

.

I_5: 'Immediately decrease the torsion load – the angular velocity is too low.'
'Immediately decrease the armature voltage.'

I_6: 'Do not increase the torsion load, but if unavoidable then increase the armature voltage.'

.

.

.

I_{15}: 'Immediately decrease the armature voltage.'

13
Modelling, simulation and expert systems

Processes and plants are designed in a rational way by avoiding dangerous situations and allowing construction and automatic actions to prevent and decrease the time of approaching or being in the dangerous zone of operation. Therefore, the probability of exhibiting a certain process state is proportional to the propensity of the plant designer and process operator to use the equipment rationally.

13.1 SEMANTICS OF THE PROCESS MODEL

Modelling processes or process parts is a very valuable and essentially good method for many procedure analyses and syntheses both from the practical and theoretical aspects. Such models are appropriate for simulations, feed-forward control and predictions. Roughly speaking there exist macro- and/or micro-models where the distinction comes from the global dimension of approach and expected results of the modelling. Modelling large-scale or compex processes cannot be performed with micro-models because of the lack of statistically relevant data at any given time. Macro-models require global complex variables while micro-models operate satisfactorily with simple process variables.

The main benefit of models is the possibility of quantitative evaluation of relevant process parameters relating to internal input and output states. Disturbing influences and parameter changes can be adequately treated as well. There is only one main course in modelling – the problem of model building, evaluation and use. The aim of model building in our case is more specific, since it has previously been used for the purpose of semantic analysis of the process or plant. When different process variables are continually observed, then the probability of occupation of a certain state is given by the probability distribution function. As detailed in Figure 13.1 one can distinguish among (1) a flat spectrum of variable prob-

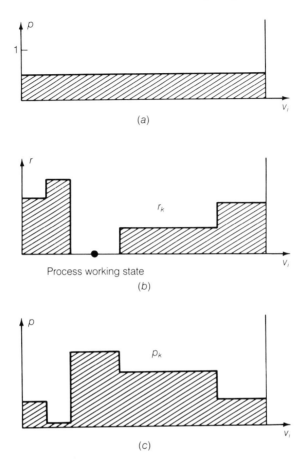

Figure 13.1 Probability distribution function and expert estimated importance levels of process variables. (*a*) flat probability distribution function; (*b*) probability distribution function measured; (*c*) importance levels of the variable estimated by an expert.

ability states, (2) the actual distribution function of the process variable, and (3) the expert estimation of importance levels of the variable amount. While the first two are a matter of actual variable changes, the third function qualitatively reflects the inversion of the second function pointing to the principal relation between probability and importance as

$$r_{jk} :: \frac{1}{p_{jk}} \tag{13.1}$$

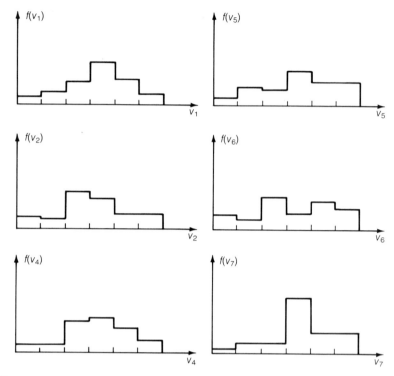

Figure 13.2 Probability distribution functions measured on the thermal power plant.

where :: stands for proportionality.

Example
The probability distribution functions of a steam-generating power plant of a mixed type – where the steam is being used both for heating and the turbine generator – would require the measuring and supervision of about 30 variables. Based on the measurement and supervision system a set of seven variables was chosen and their relations stated in a macro-model of daily cycles of steam generation as

$$a_0 + a_1 v_1 + \frac{a_2 v_2}{a_3 + a_4 v_3} + a_5 v_4 + a_6 \frac{v_5 v_6}{v_7} = 0 \qquad (13.2)$$

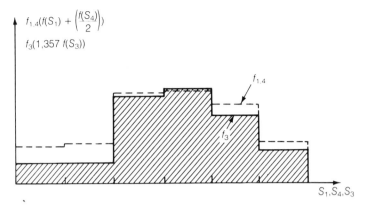

Figure 13.3 The similarity of probability distribution functions of complex variables for a thermal power plant.

where

v_1 is the feedwater flow/intake (t/h),
v_2 is the fuel flow intake (l/h) – masoot,
v_3 is the fuel temperature (°C),
v_4 is the superheated steam flow (t/h),
v_5 is the temperature of exhausted gases (°C),
v_6 is the burner air flow/intake (Nm^3/h),
v_7 is the differential pressure of air for burners (mm of water)

and

$a_0 - a_6$ are model constants.

The model building is essentially based on the non-linear method of rank correlation which forces the importance factor of particular variables, or groups the effects of several variables in the total balance of variable influences on the process state. The probability distribution functions of variables in Figure 13.2 are given as registered from the process. Taking the last four parts of expression (13.2) as macro variables, we can obtain the relation

$$a_0 + S_1 + S_2 + S_3 + S_4 = 0 \qquad (13.3)$$

where $S_1 - S_4$ are complex process variables. The complex variable S_4 equals

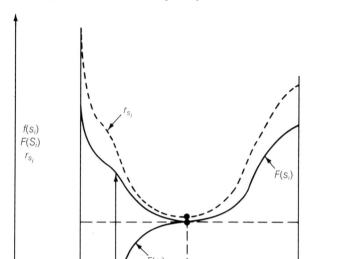

Figure 13.4 Graphical construction of the importance level function from distribution function of the complex process variable S_i.

$$S_4 = a_6 v_5 v_6 / v_7,$$ (13.4)

By linear transformation of probability distribution functions of process variables (Papoulis, 1976) as shown in Figure 13.3 we can obtain a satisfactory quantitative relation between the complex process variables of the model.

The distribution functions of complex model variables can be used in many ways for the evaluation of the importance parameter of particular process states. By using a modified distribution function as shown in Figure 13.4, a practical and possible way of the evaluation of the importance parameter can be obtained. Here the expression for the quantitive estimation of the importance parameter r_i of the complex variable is equal to the weighted discrepancy regarding the process balance point, i.e.

$$r_{s_i} = w_i \left| \int_{s_{i_0}}^{s_i} f_{s_i} d_{s_i} \right|$$ (13.5)

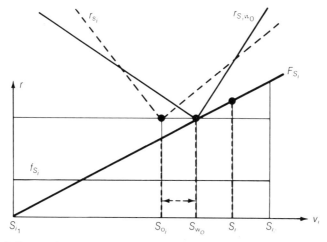

Figure 13.5 Determination of the importance parameter for an asymmetrical working state of the process.

where w_i is the weighting constant of the importance parameter, depending on the proportion of the complex variable in expression (13.3), S_{i_0} is the process balance point expressed as depending on variable S_i, S_i is the instantaneous working point, and f_{s_i} is the probability distribution function of the complex process variable S_i.

Expression (13.5) prefers the semantic estimation of the process in the middle part of the probability distribution function. When the process working set point differs from the process balance point, then the procedure in Figure 13.5 can be followed and the importance parameter of the complex process variable calculated as

$$r_{s_i w_o} = \left(\omega + k \left| \int_{s_{i_0}}^{s_{i_w}} f_{s_i} ds_i \right| \right) \times \left| \int_{s_{i_0}}^{s_i} f_{s_i} ds_i \right| \tag{13.6}$$

where S_{iw} is the set point of complex process variable S_i.

Nevertheless, models are simplified descriptions of the process and for our example seven variables were taken from the set of 30 measurable variables as sufficiently representative. Two problems were met: (1) a precise determination of the model constants $a_0 \ldots a_6$ and (2) the additive effect of the model behaviour by changes in constant a_0 influenced by changes in all other than those seven model variables. The first problem can be solved by taking the sensitivity function in the form

$$\frac{\partial S_i}{\partial V_i} = S_{enj} \qquad (13.7)$$

for the determination of the influence of variable V_j on the ith complex variable.

The compensation of the a_0 parameter change can be done by several on-line measurements of the variables during daily production cycles. Thus the importance of the process variables irrelevant to the model can be determined by process experts for each particular case and presented as separated items, while the importance parameter of model variables can be estimated by long-term measurement and determined for each working point. Since the model does not include component failure and malfunctions it cannot be used for such purposes unless an adquate knowledge base is organized.

13.2   FUZZY VARIABLES, SIMULATION AND DECISION MAKING

When complex processes are being reduced to several key variables there comes a point when descriptive variable parameters should be used to improve the process description. The distinction between small and big and true and false are not given in strongly defined limits for such variable descriptions and attributes (Zadeh, 1975). They were named 'fuzzy' variables and theoretically described as fuzzy set theory. One of the application fields of fuzzy set theory is process modelling and decision making. A fuzzy relation R between two fuzzy sets A (input) and B (output) can be defined in the following manner (Bellman and Zadeh, 1970)

$$R = (A \times B) \cup (\bar{A} \times U) \qquad (13.8)$$

where $A \times B$ is a Carthesian product over fuzzy sets A and B; $\bar{A}$ is the complement of the set A and U is the universal set. By using the membership function description the relation R equals

$$R = (\mu_A \cap \mu_B) \cup (\mu_{\bar{A}} \cap \mu_U) \qquad (13.9)$$

where μ_A, μ_B, $\mu_{\bar{A}}$ and μ_U are the membership functions of sets A, B, $\bar{A}$ and U, and $\cap$ and $\cup$ are min and max operators. Many different fuzzy set relations have been proposed, based on the principle of logical reasoning but no single one satisfies all the criteria of reasoning (Averkin *et al.*, 1986). Nevertheless, by using fuzzy set relations the presentation of the fuzzy decision making process is possible in the following way.

Over the universe of fuzzy sets X, Y, Z let us define fuzzy sets $A_1, \ldots, A_n$, $B_1, \ldots, B_n$ and $C_1, \ldots, C_n$. Propositions between these sets of the type

If A_1 then (if B_1 then C_1)

or

If A_2 then (if B_2 then C_2)

or

$\vdots$

or

If A_n then (if B_n then C_n)   (13.10)

define the relation $R(X, Y, Z)$. As can be seen, all the rules are naturally equal and each rule will give a certain decision C_i for some starting values from X and Y, so that the final decision can be represented as a set of individual decisions

$C_1, C_2, \ldots, C_n$ (13.11)

The final decision equals

$C = \text{Max}(C_1, C_2, \ldots, C_n)$ (13.12)

or as described with the membership function for n rules

$\mu(X, Y, Z) = \text{Max}\mu_{R_i}(x, y, z), i = 1, \ldots, n$ (13.13)

As this decision model is simple, understandable and acceptable, it has been applied for process modelling and process control decision making. In spite of its satisfactory initial results it has not found an adequate application in practice. The reason for this lies in the following:

1. Lack of real hierarchy in the decision process; all the rules are mutually equal and possible hence they should all be examined prior to the optimum decision selection.
2. For complex systems, especially for many input – many output

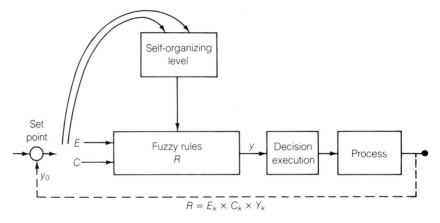

$$R = E_k \times C_k \times Y_k$$

Figure 13.6 The structure of the fuzzy self-organizing controller. Rule: if error is *n*egative *m*edium then if change in error is *p*ositive *s*mall then the deviation of the response is *p*ositive *s*mall. *R*: NM × PS ⇒ PS; *E*: error; *C*: change in error.

Table 13.1 A simplified case of performance measure rules *R*

| *C* | *E* | | | | | | | |
---	PL	PM	PS	PO	NO	NS	NM	NL
PL	NL	NL	NL	NM	O	O	O	O
PM	NL	NL	NM	NS	O	O	O	O
PS	NL	NM	NS	O	O	O	PS	O
O	NL	NM	NS	O	O	PS	PM	PL
NS	O	NS	O	O	O	PS	PM	PL
NM	O	O	O	O	PS	PM	PL	PL
NL	O	O	O	O	PM	PL	PL	PL

P: Positive   M: Medium   E: Error
N: Negative   S: Small   C: Change in error
L: Large

systems, the number of rules can be too large to allow the decision making process in a given time margin.

3. Presentation of fuzzy relations in the controller memory as a relation matrix may demand a significant memory content. By taking the number of universal set attributes as no fewer than 15, then for one input or output system the relation matrix

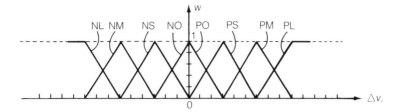

Figure 13.7 The most usual method of membership function definition. NL: negative large; NM: negative medium; NS: negative small; NO: negative zero; PO: positive zero; PS: positive small; PM: positive medium; PL: positive large; w: fuzzy weighting factor of the membership function; Δv_i: variable error/difference.

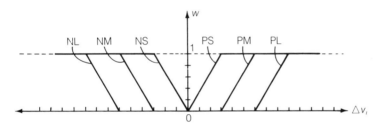

Figure 13.8 A non-standard definition of the membership function (Marković, 1990).

demands about 1000 elements, while for two input–two output systems a relation matrix has about one million elements (Shao, 1988).

Nevertheless, by using fuzzy set theory one can obtain, for example, the controller structure given in Figure 13.6 and the performance measure of the linguistic type as given in Table 13.1. This simplified table is actually defuzzified (since there are no overlapping elements), unchangeable and qualitative (since there are no measures of membership). This table also lacks any probability measure.

Principally, all the defects come from the fact that there are no limits to the definition of the membership function of the fuzzy set. In the practice of process control, the most frequent way of defining the membership function of fuzzy sets is that given in Figure 13.7. A more refined way of defining the membership function with restriction is given in Figure 13.8 (Marković, 1990) or mathematically defined as

(1)

$$\mu_{x_i} = \begin{cases} 0 & \text{for } \mu_{x_{i-1}} = [0, 1) \\ [0, 1] & \text{for } \mu_{x_{i-1}} = 1 \quad \text{and} \quad \mu_{x_{i+1}} = 0 \\ 1 & \text{for } \mu_{x_{i+1}} = (0, 1] \end{cases} \tag{13.14}$$

(2) μ_{x_i} is a monotonically increasing function.

Such a fuzzy decision model allows:

1. hierarchical propositional rules whereby the decision (and action) is possible with only one pass through the decision tree, with mutually exclusive proposition rules;
2. inclusion of the probability function into fuzziness, since the membership function given in equation (13.14) has all the properties of the probability function;
3. reduction of the required memory space, since there is only one membership function changing in the given parameter space;
4. a probability of fast learning, since once established the membership function can be changed with changing process parameters.

13.3 PROCESS DECISION MODELLING – THE CASE OF A POWER PLANT

The simulated process is a power plant of mixed type, i.e. simultaneously producing heating steam and supplying steam to a turbine for electrical power production. The plant is controlled by classical analogue PID controllers and supported by a digital event recorder for about 400 binary and 200 analogue variables. The scanning cycle of the event recorder was 0.5 s for all registered values. The simulated process is presented schematically in Figure 13.9. The goal of the controller is to maintain a constant pressure of superheated steam at the output. There are four burners that can be integrally controlled and represented as a single unit. The steam line from the boiler to the turbine is not provided with a one direction control valve which would preserve a constant pressure at the boiler output, but the pressure varies with the change of the load. Therefore, the maintenance of constant pressure at the boiler output and the return of the pressure to the required limits is the primary task of the automation equipment and crew.

The simulated strategy of the boiler control is formulated in discussion with the operators, production technologists and the I&C manager. The change of pressure of the superheated steam at the

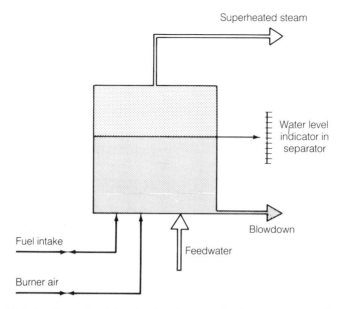

Figure 13.9 A schematic of the simulated process in the steam generating unit.

boiler output compared to the required value of 116 bar requires the intervention of operators primarily relating to the fuel consumption and burner air intake and then to the feedwater intake. Regarding the fact that the observed process is relatively slow, i.e. from the operator action to the change of output pressure there is a time interval, the amount of the change of fuel and air consumption is dependent on two factors

1. the amount of the pressure change;
2. the instantaneous fuel intake.

The changes in the fuel intake were described by the operators as linearly dependent on changes of pressure and fuel intake. The system actions regarding pressure changes were separated into actions changing the fuel and air intake and changes in the boiler water level. Thus, the input fuzzy set $A_1, \ldots, A_n$ was defined according to Figure 13.10 as the pressure change at the power plant output. As a final consequence of the pressure change comes the change in the fuel intake burner air and feedwater consumption. The

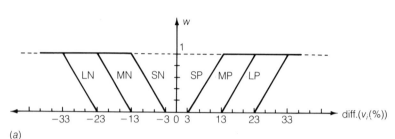

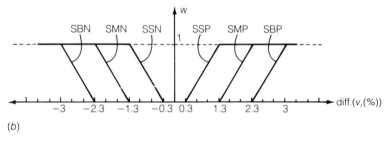

Figure 13.10 (*a*) Membership functions of the input fuzzy set for coarse control of the thermal power plant. (*b*) Membership functions of the input fuzzy set for fine control of the thermal power plant. SBN: small big negative; SMN: small medium negative; SSN: small small negative; SSP: small small positive; SMP: small medium positive; SBP: small big positive.

feedwater intake depends both on steam production and desalination. Therefore, the feedwater intake is connected to the water level in the boiler and to water for desalination and represented as a separate entity.

The burner air flow is directly (linearly) connected to the fuel consumption with the constant of $10.1\,m^3$ burner air for one litre of masoot. In this way there were defined output fuzzy sets connected to the variable of the fuel intake, described in Figure 13.11. Generally, the decision action at the output Y depends on the input difference E and the input change of difference C connected with the system matrix R in the following manner

$$Y = (E \times C)R \tag{13.15}$$

By using the input and output sets as defined in Figures 13.11 and 13.12 the following relations hold

$$A_n < A_{n-1} < \ldots < A_1$$
$$B_n < B_{n-1} < \ldots < B_1 \tag{13.16}$$

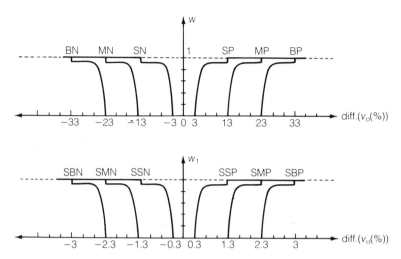

Figure 13.11 Membership functions of the output fuzzy set for coarse and fine control of the thermal power plant. diff(v_0 (%)): percentage difference in the output variable in successive time intervals.

for both positive and negative changes in the input steam pressure and output (fuel intake). By using relation (13.3), neglecting a_0 and putting real values of complex variables S_1, S_2, S_3 and S_4, the relative influence of the variables is that given in relation (13.17). Therefore, the fuel intake was taken as the main response variable.

$$
\begin{array}{ccccc}
\text{(feedwater)} & \text{(fuel/air)} & \text{(steam)} & \text{(exhaust gases)} & \\
S_1 & + \quad S_2 & + \quad S_3 & + \quad S_4 & = \text{TOTAL} \\
18.21\% & 31.79\% & 27.76\% & 22.74\% &
\end{array}
$$

$$(13.17)$$

The main hierarchical variable is the pressure change in two successive time intervals. Table 13.2 presents actual levels of the main hierachical variables. The relation R is given as a series of propositional rules and presented graphically in Figure 13.12 for the positive pressure change.

Supposing the pressure change to be +11 bar between the last two samples, there are three possibilities:

1. a new steam consumer was switched on;
2. increased fuel in the last interval has not yet had an effect;
3. a previous decrease of the fuel intake was too large.

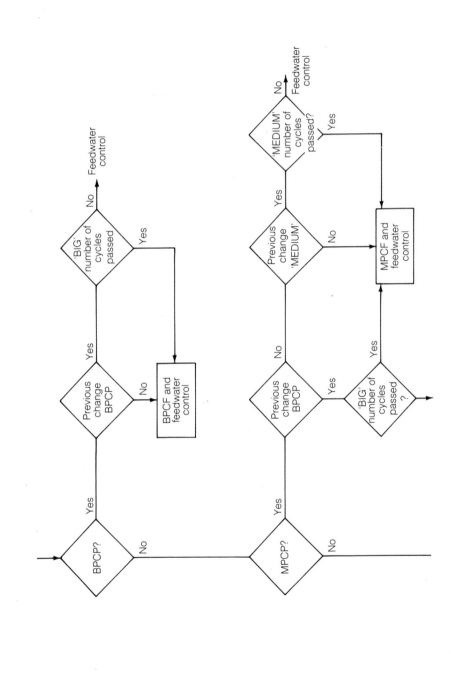

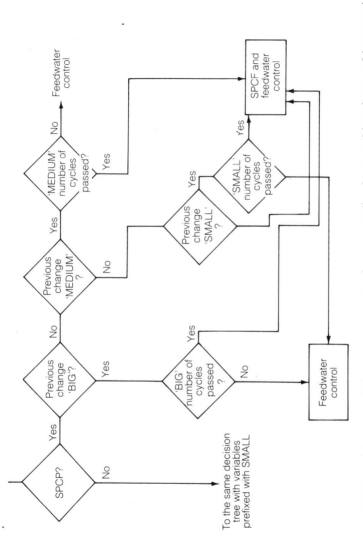

Figure 13.12 Decision tree of the steam generator fuzzy controller for a positive change of input variable (Marković, 1990).

Table 13.2 Actual levels of hierarchical process variable changes

Type of change	Change of input variable	Acronym for positive change
Big change of fuel intake	–	BPCF
Medium change of fuel intake	–	MPCF
Small change of fuel intake	–	SPCF
Big change of pressure	>23 bar	BPCP
Medium change of pressure	>13 bar	MPCP
Small change of pressure	> 3 bar	SPCP
Small big change of fuel intake	–	SBPCF
Small medium change of fuel intake	–	SMPCF
Small small change of fuel intake	–	SSPCF
Small big change of pressure	>2 bar	SBPCP
Small medium change of pressure	>1 bar	SMPCP
Small small change of pressure	>0.1 bar	SSPCP

The program should ask for two facts concerning the system behaviour in previous intervals:

1. whether the last increase was large, medium or small, and
2. how much time has elapsed since the previous action.

Supposing the last increase was small and that the action time interval has just passed then a 'small' increase in the fuel intake should take place. The amount of the increase is determined by a min/max composition of input and output sets. The decision is thus made and no further investigation of rules is necessary. The decision path for this case is given in Figure 13.12. The consequences of particular decisions last a different amount of time, the largest relating to the greatest change of load.

14
Extracting knowledge

Knowledge in process control consists of deterministic and stochastic components. The main deterministic parts are:

1. the process component description, including its interconnections;
2. deterministic rules that connect process variables and required actions;
3. a fixed prescription for the operator behaviour in certain process situations.

The stochastic components are:

1. the probabilities of process events – facts;
2. prior probabilities of the given hypotheses;
3. the attribute distribution of variables and facts and its (semantic) evaluation;
4. stochastic rules connecting hypotheses, facts, events and attributes.

When a certain stochastic rule dominates in a given situation it automatically becomes more or less deterministic. There is no line that separates these two fields, but merely experience which, however, leads to and stays on the ground of individuality. On the other hand the performance of each decision or action can be measured unless obscured by noise or overlapped by another action.

14.1 AUTOMATIC KNOWLEDGE ACQUISITION

A process changes its state spontaneously or under control influences or actions. Eight states of the water tank from Figure 5.3 were recognized according to the positions of binary switches FS1, FS2 and PS. The state transition matrix TM given in Table 14.1 fully describes all possible state exchanges. The state transition prob-

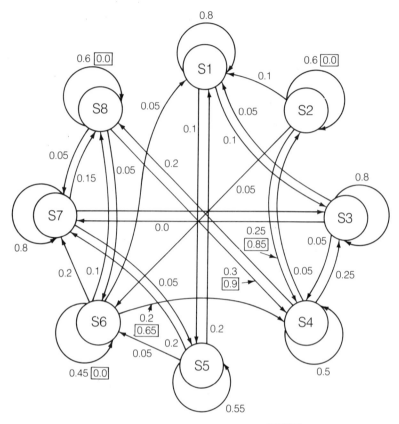

Figure 14.1 Transition graph for data in Table 14.1. $\boxed{\text{x.xx}}$ denotes the change in probability induced by automatic or operator action.

abilities are given as matrix elements and present the possible direction of change among states in succesive time intervals of the system observation. Thus the sum of probabilities in each row equals one. The corresponding transition graph is given in Figure 14.1. The present state depends on the first state that occurs, but a prior probability of any state occurring can be calculated by simply multiplying several times transition matrix by itself (Hudoklin, 1976). The necessary and sufficient conditions for this procedure are:

1. inseparability of the transition matrix into sub-matrices;
2. a non-oscillatory transition matrix, i.e. the impossibility of the return to any state from any state with a probability of one.

Table 14.1 Process states and transition probabilities for the process given in Figure 4.4

Process state code			Process state at	Process state at moment t_1							
PS1	BS2	FS	$t_1 - t$	S_1	S_2	S_3	S_4	S_5	S_6	S_7	S_8
0	0	0	S_1	0.8	0 *0.0*	0.1	– *0.85*	0.1	–	–	–
0	0	1	S_2	0.1	0.6	–	0.25	–	0.05	–	–
0	1	0	S_3	0.05	–	0.8	0.05	–	–	0.1	–
0	1	1	S_4	–	0.05	0.25	0.5	–	–	–	0.2
1	0	0	S_5	0.2	–	–	– *0.65*	0.55	0.05 *0.00*	0.2	–
1	0	1	S_6	0.05	–	–	0.2	–	0.45	0.2	0.1
1	1	0	S_7	–	–	0.05	– *0.9*	0.05	–	0.8	0.15 *0.0*
1	1	1	S_8	–	–	–	0.3	–	0.05	0.05	0.6

Transition – means non-applicable transition
Transition probabilities changed by automatic/operator's action are in italic

Control or operator actions change the process state and usually the operator should prefer transitions from dangerous to safe states as indicated in Table 14.1. This also changes the matrix elements of probabilities and prior probabilities. Such transitions can even change the nature of the matrix itself, making it separable or oscillatory and preventing the determination of prior probabilities. Nevertheless, when these transitions are observed through a surveillance system, the prior probabilities can be automatically collected. Automatic collection of hypothesis probabilities is connected to a much more complex analytical and data acquisition procedure where the essence lies in the opposite tendency to make the deterministic world more stochastic in order to avoid details. As an example let us consider the hydromechanical and integral substitutional scheme of a hydromechanical corn sprayer as given in Figures 14.2 and 14.3 (Srabotnak, 1989). A substitutional stationary scheme can be obtained by using the following facts and approximations:

1. the influence of pressure and flow oscillations on the reciprocating pump were neglected;
2. the mean flow through the membrane amortizer equals zero, thus excluding this element from the scheme;

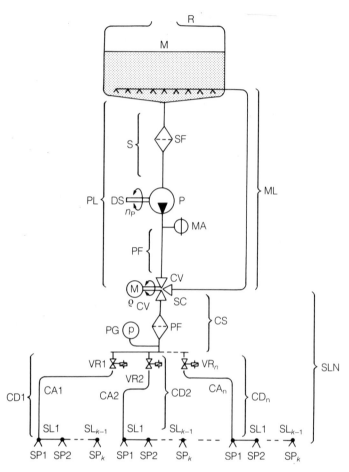

Figure 14.2 Hydromechanical scheme of a corn sprayer: the components are: reservoir R, suction filter SF, suction flue S, pump P, membrane amortizer MA, driving shaft DS, pressure flue PF, mixture side of the control valve CV, mixture flue M, spraying side of the control valve SC, pressure filter PF, circuit valves VR1, VR2, . . . ,VRn, conduits CA1, CA2, . . . , CAn, spraying lines SL1, SL2, . . . , SL$_{k-1}$, sprayers SP1, SP2, . . . , SPk, pump line PL, sprayer line SLN, mixing line ML, conduits from distribution valves CD1, CD2, . . . , CDn, pressure gauge PG and common sprayer flue CS (Srabotnak, 1989).

3. the time constant of the reservoir discharge is so big that it can be considered as a pure pressure source;
4. conduit inertia is negligible, thus making conduits pure resistance elements;
5. spraying lines SL_x have a negligible resistance, making a parallel combination of each spraying segment;
6. resistances of CD_X (conduits and distribution valves), although with different length of lines, are approximately of the same value as the pressure p_{CD} and flow $q_{mln/nk}$

Such a scheme can lead to a simplified stationary and dynamic scheme, as given in Figures 14.4(*a*) and (*b*). In order to recognize the influence of each spraying element change on the functioning of the pump, one has to take into account the influence of change on both stationary and dynamic characteristics. This can lead to the same changes influenced by several causes and to the introduction of different hypotheses. Such hypotheses can be supported only by prior probabilities of their appearance. Nevertheless, small differences in the final effects on sprayer characteristic due to different causes of their appearance cannot be detected by any hypothesis. They have to be tested against all possible hypotheses. An analytical relation that can be obtained from the presented schemes can help in the determination of the prior probabilities of hypotheses. By using on-line data, an automatic change of prior probabilities occurs, as given in Chapter 7.

14.2 SOME LIMITS ON DATA PROCESSING FOR KNOWLEDGE BASES

Knowledge is needed for a proper process action. Knowledge bases have to be supplied with additional software, e.g.

1. a protection algorithm for process actions;
2. process modelling for hypothesis support;
3. process simulation for semantic evaluations;
4. a performance function for the process action evaluation.

The performance function applied to fuzzy process variables is per-

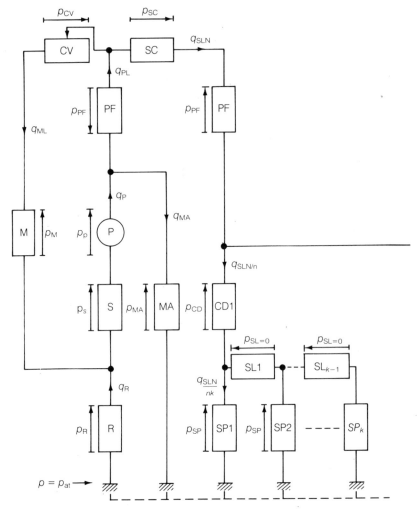

Figure 14.3 Integral substitutional scheme of the hydromechanical part of the corn sprayer (abbreviations given in Figure 14.2). q = flow, p = pressure.

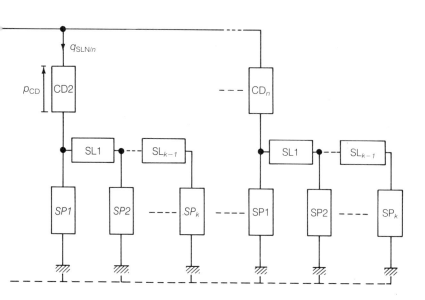

Figure 14.3 Continued

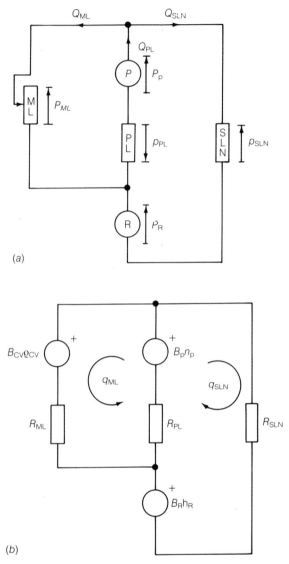

Figure 14.4 (*a*) Substitutional scheme of the stationary state of corn sprayer. (*b*) Substitutional dynamic scheme of corn sprayer; $B_{CV}\rho_{CV}$, $B_p n_p$ and $B_R h_R$ are the active pressures of control valve, pump and reservoir.

formed in the way described in Table 14.2 and Figure 14.5. It shows the effects of action after the expected time interval, whereas the time interval differs for the small, middle or large change in the output variable. Its functioning is possible only when there is no disturbance evidenced within the observed time interval. The limit of

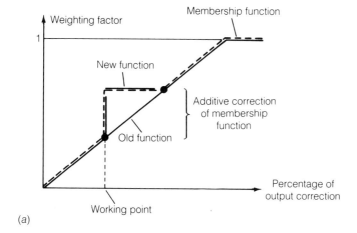

(a)

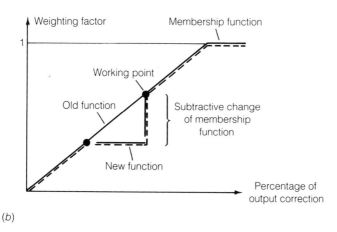

(b)

Figure 14.5 Corrective changes in fuzzy set membership function due to action performance. (*a*) additive change of the membership function, (*b*) subtractive change of the membership function.

Table 14.2 Change of output membership function caused by inadequate transition actions in previous time interval

Transition type after elopsed time interval	Reason for transition	Change of membership function of output variable
Output variable increase – output variable increase	Inadequate first action	Additive change for each member lower than the sum of correction (Figure 14.5(*a*))
Output variable increase – stationary state	No reason for change	No change of membership function
Output variable increase – output variable decrease	Too big first action	Subtractive change for each member lower than the difference of correction (Figure 14.5(*b*))

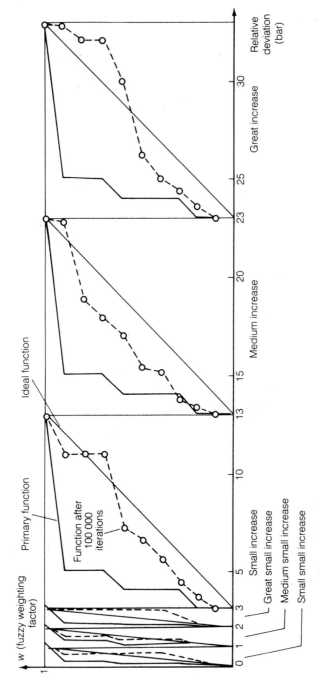

Figure 14.6 Fuzzy variable and primary learning results – the case of a power plant.

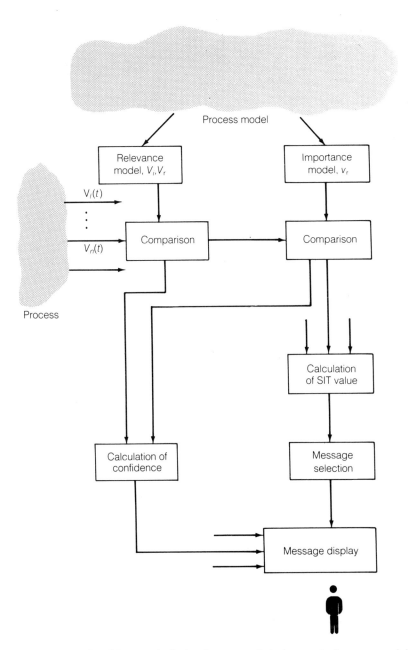

Figure 14.7 Confidence calculation for semantic judgement of process variables.

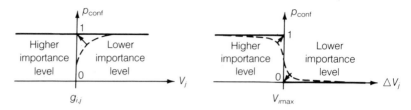

Figure 14.8 Two confidence functions improved by long-term semantic judgement.

the action performance is given within defined input and output variables and its changes, i.e. in the event of a structural change in the process new variables and fuzzy sets have to be defined. The result of the learning procedure for the case of a power plant is given in Figure 14.6 after 100000 learning iterations. Due to a fast and simple algorithm (Figure 13.13) these 100000 iterations were performed in five minutes on a 0.7 MIPS machine giving the minimum time response of the method (Markovic and Jović, 1991).

14.3 OPERATIONALIZATION AND LIMITS OF LEARNING

A message generated in the controller is based on the particular SIT value and connected to the actual process state. This semantic value is calculated on the basis of the relevance and importance of particular process variables. In the case of the existence of the description of all process states needed for a qualitative surveillance of the process, distrust about the regularity of the selected message at the message base falls on the regularity of the relevance and importance judgement of particular process variables.

The relevance parameter is judged by a variable comparison in two successive samples with the modelled difference and with the modelled amount of the process variable defined by an expert or by the process model. The confidence in the process message will therefore be lower when the variable is on the verge of relevance or within two importance levels. A total confidence in the judgement of the process variable $V_i(t)$ is based on the confidence in the relevance parameter and confidence in the importance parameter. Figure 14.7 describes the procedure of the confidence calculation of process variables. Depending on the process nature, an iterative improved process of semantic judgement can be established where the confidence function P_{ir} and P_{is} from Figure 14.7 can become idealized as given in Figure 14.8.

REFERENCES for Part Four

Ahrens, W. (1987) *Automatisierungstechnische Praxis.* **29**, **10**, 475–84.

Averkin, A.N. *et al.* (1986) *Fuzzy sets in control models and artificial intelligence* (in Russian). Nauka, Moscow.

Bartos, F.J. (1989) *Control Engineering.* July, 90–3.

Bellman, R.E. and Zadeh, L.A. (1970) *Management Science.* **17**, B-141–B-164.

Bieker, B. (1987) *Automatisierungstechnische Praxis.* **29**, **1**, 36–43.

Fogarty, T.C. (1989) *Proceedings of the fifth Conference on Artificial Intelligence Applications, Miami, Florida.* pp. 215–21.

Gitt, W. (1986) *Energie—optimal durch die Information.* Hanssler, Neuhausen – Stuttgart.

Hudoklin-Božič, A. (1986) Stohastični procesi, Universıty in Maribur, VŠOD Kranj (in Slovenian).

Leung, K.S. and Lam, W. (1988) *Computer.* September, 43–56.

Mamdani, E.H. (1974) *Proc. IEEE.* **121**, 1585–8.

Marković, P. (1990) *Fuzzy set hierarchical model with variable structure.* Ph.D. thesis, Faculty of Electrotechnics, University of Zagreb.

Marković, P. and Jovic, F. (1991) A hierarchical fuzzy set decision model, *Automatika*, **3**, **4**, (32).

Papoulis, A. (1976) *Probability, random variables and stochastic processes.* McGraw-Hill, New York.

Shao, S. (1988) *Fuzzy sets and systems 26.* pp. 151–64. North-Holland, Amsterdam.

Srabotnak, F. (1989) Personal Communication.

Wilhelm, R. (1979) *IEEE Trans. Aut. Control* **24**, 1–27.

Witting, T. (1987) *Technische Mitteilungen Krupp* **2**, 71–8.

Zadeh, L.A. (1975) The concept of a linguistic variable and its application to approximate reasoning, *Information Sciences* **8**, **9**, 199–251, 301–57, 43–80.

Zupanec, R. (1989) *Process state surveillance using qualitative estimation.* M.Sc. Thesis, Faculty of Electrotechnics, University of Zagreb.

Appendix – List of expert system firms and institutes

1 Asea Brown Boveri AG
  Corporate Research Heidelberg
  PO Box 101332
  D-6900 Heidelberg, Germany
2 Aion
  101 University Avenue
  Palo Alto, CA 94301
  USA
3 AXION A/S (AXI)
  Bregnerøfvej 133
  DK-3460 Birkerød
  Denmark
4 CGS Institute
  Russell House, Russell Street
  Windsor, Berkshire SL4 1HQ
  Great Britain
5 Dornier GmbH
  Postfach 1420
  D-7990 Friedrichshafen 1
  Germany
6 ecn Neurocomputing GmbH
  Max-von-Eyth-Strasse 3
  8045 Ismaning
  Germany
7 Epsilon
  Kurfürstendamm 188/189
  D-1000 Berlin 15
  Germany
8 First Class Expert Systems
  286 Boston Post Road
  Wayland, MA 01778
  USA

9 Ferranti GmbH
   Taunusstr. 52
   6200 Wiesbaden
   Germamy
10 Fraunhofer Institut für Transporttechnik und Warendistribution
   Emil-Figge-Strasse 75
   4600 Dortmund 50, Germany
11 Gepard
   Am Kanal 27
   1110 Wien
   Austria
12 Gold Hill Computers
   26 Landsdowne Street
   Cambridge, MA 02139
   USA
13 Hahn-Meitner-Insitut Berlin GmbH
   Glienecker Strasse 100
   D-1000 Berlin 39
   Germany
14 Hema
   Röntgenstr. 31
   7080 Aalen
   Germany
15 Information Futures Limited
   Crays Pond House, Crays Pond,
   Reading, Berkshire RG8 7QG
   Great Britain
16 Intelligent Technology Group
   115 Evergreen Heights Drive
   Pittsburgh, PA 15229
   USA
17 ISRA Systemtechnik GmbH
   Mornewegstrasse 45A
   6100 Darmstadt
   Germany
18 Katholieke Universiteit Leuven
   Chemical Engineering Department
   de Croylaan 46
   Belgium
19 Krupp Atlas Elektronik GmbH
   Sebaldsbrücker Heerstr. 235
   D 2800 Bremen 44
   Germany

20 Logic Programming Associates Ltd
    Studio 4, Royal Victoria Patriotic
    Building, Trinity Road,
    London SW18 3SX, Great Britain
21 MSC
    Werner v. Siemensstr. 1
    7513 Stutensee 3
    Germany
22 Neural Ware Inc
    103 Buckskin Court
    Sewikley, PA 15143
    USA
23 Neuron Data
    444 High Street
    Palo Alto, CA 94301
    USA
24 Perception International
    211 West Mountain Road,
    Ridgefield
    Connecticut 06877, USA
25 PrediSys Ltd.
    Vapaalantie 2 B Postf. 68
    Vantaa
    Finnland
26 Sira Ltd
    South Hill, Chislehurst,
    Kent BR7 5EH
    Great Britain
27 University Duisburg
    Dept. for Machine Engineering
    Lotharstr. 1–21
    4100 Duisburg 1
    Germany
28 University of Strathclyde
    Scottish HCI Centre
    George House, 36 North
    Hanover Street
    Glasgow G1 2AD, Great Britain
29 THORN EMI Software TECS
    Hanover House, Plane Tree
    Crescent, Lower Feltham,
    Middlesex TW13 7AQ
    Great Britain

Index